Springer Theses

Recognizing Outstanding Ph.D. Research

Aims and Scope

The series "Springer Theses" brings together a selection of the very best Ph.D. theses from around the world and across the physical sciences. Nominated and endorsed by two recognized specialists, each published volume has been selected for its scientific excellence and the high impact of its contents for the pertinent field of research. For greater accessibility to nonspecialists, the published versions include an extended introduction, as well as a foreword by the student's supervisor explaining the special relevance of the work for the field. As a whole, the series will provide a valuable resource both for newcomers to the research fields described, and for other scientists seeking detailed background information on special questions. Finally, it provides an accredited documentation of the valuable contributions made by today's younger generation of scientists.

Theses are accepted into the series by invited nomination only and must fulfill all of the following criteria

- They must be written in good English.
- The topic should fall within the confines of Chemistry, Physics, Earth Sciences, Engineering, and related interdisciplinary fields, such as Materials, Nanoscience, Chemical Engineering, Complex Systems and Biophysics.
- The work reported in the thesis must represent a significant scientific advance.
- If the thesis includes previously published material, permission to reproduce this must be gained from the respective copyright holder.
- They must have been examined and passed during the 12 months prior to nomination.
- Each thesis should include a foreword by the supervisor outlining the significance of its content.
- The theses should have a clearly defined structure including an introduction accessible to scientists not expert in that particular field.

More information about this series at http://www.springer.com/series/8790

Kohei Kusada

Creation of New Metal Nanoparticles and Their Hydrogen-Storage and Catalytic Properties

Doctoral Thesis accepted by Kyoto University, Kyoto, Japan

Author
Dr. Kohei Kusada
Kyoto University
Kyoto
Japan

Supervisor
Prof. Hiroshi Kitagawa
Kyoto University
Kyoto
Japan

ISSN 2190-5053 ISSN 2190-5061 (electronic)
ISBN 978-4-431-56406-5 ISBN 978-4-431-55087-7 (eBook)
DOI 10.1007/978-4-431-55087-7

Springer Tokyo Heidelberg New York Dordrecht London

Printed on acid-free paper

Springer is part of Springer Science+Business Media (www.springer.com)

Parts of this thesis have been published in the following journal articles:

- Kusada K, Yamauchi M, Kobayashi H, Kitagawa H, Kubota Y (2010) Hydrogen-Storage Properties of Solid-Solution Alloys of Immiscible Neighboring Elements with Pd. *J. Am. Chem. Soc.*, 132:15896–15898.
- Kusada K, Kobayashi H, Ikeda R, Morita H, Kitagawa H (2013) Changeover of the Thermodynamic Behavior for Hydrogen Storage in Rh with Increasing Nanoparticle Size. *Chem. Lett.*, 42:55–56 (Editor's Choice).
- Kusada K, Kobayashi H, Yamamoto T, Matsumura S, Sumi N, Sato K, Nagaoka K, Kubota Y, Kitagawa H (2013) Discovery of Face-centered Cubic Ruthenium Nanoparticles: Facile Size-controlled Synthesis using the Chemical Reduction Method. *J. Am. Chem. Soc.*, 135:5493–5496
- Kusada K, Kobayashi H, Ikeda R, Kuboata Y, Takata M, Toh S, Yamamoto T, Matsumura S, Sumi N, Sato K, Nagaoka K, Kitagawa H (2014) Solid Solution Alloy Nanoparticles of Immiscible Pd and Ru Elements Neighboring on Rh: Changeover of the Thermodynamic Behavior for Hydrogen Storage and Enhanced CO-Oxidizing Ability. *J. Am. Chem. Soc.*, 136:1864–1871

Supervisor's Foreword

My department nominated this Ph.D. thesis for a Springer Thesis because I regard it as a remarkable work carried out with a quite high level of independence. Kohei Kusada joined my group in 2007 as an undergraduate student. My research group has been working on metal nanoparticles having unique structure and properties. Kohei had studied the creation of novel metal nanoparticles having hydrogen absorption properties and/or catalytic activity. His synthetic method is notable for its facility and flexibility, and this method can be used to obtain metal nanoparticles having nonequilibrium structures that do not exist in bulk state. Also the novel nanoparticles that he synthesized provide a new strategy on the basis of inter-elemental fusion to create highly efficient functional materials for energy and material conversions.

This thesis reports four central themes of metal nanoparticles. First, the synthesis method of AgRh nanoparticles is established and its hydrogen-storage property is observed. Second, high catalytic activity for CO oxidation using novel PdRu nanoparticles is investigated. Third, the structure of Ru nanoparticles (hcp and fcc lattices) is controlled by facile synthetic methods. Fourth, the hydrogen-storage properties of Rh in the boundary region between nanoscale and bulk are investigated. Some of these interesting results were published as a full-length research article or a research letter in the *Journal of the American Chemical Society*, and their novelty and significance have been highly evaluated by many peers. In addition, his results have been featured in a number of conferences. These are proofs that this thesis is worthy of a Springer Thesis.

Kyoto, Japan, April 2014 Prof. Hiroshi Kitagawa

Acknowledgments

This thesis is the summary of my studies from April 2007 to December 2012 under my supervisor, Prof. Hiroshi Kitagawa, in the Division of Chemistry, Graduate School of Science, Kyoto University, and the Department of Chemistry, Faculty of Science, Kyushu University.

My heartfelt appreciation goes to Prof. Kitagawa, whose enthusiastic tutelage and generous suggestions were of inestimable value for my study and expanded my view of the world. He gave me many unprecedented opportunities to meet all kinds of extraordinary people, and to learn about not only the leading-edge research but also foreign culture. There is no doubt that all of the study involved in this thesis could not have been performed without his mentorship.

I am sincerely grateful to Prof. Miho Yamauchi at Kyushu University and to Prof. Hirokazu Kobayashi for their warm-hearted encouragement, valuable discussions, and courteous instructions. Especially when I was a Bachelor's degree student, their considerate guidance led me to pursue a career as a researcher.

Prof. Yoshiki Kubota at Osaka Prefecture University gave me insightful comments and suggestions about X-ray diffraction measurements at SPring-8. Discussions with Prof. Syo Matsumura, Dr. Shoichi Tho, and Mr. Tomokazu Yamamoto at Kyushu University illuminated the structure of samples through TEM measurement.

I have had great support in experiments and in discussions with Prof. Katsutoshi Nagaoka, Dr. Katsutoshi Sato, and Mr. Naoya Sumi at Oita University. Affiliate Prof. Ryuichi Ikeda gave me constructive comments and warm encouragement.

I am deeply grateful to Prof. Cameron J. Kepert at the University of Sydney for giving me the opportunity for education abroad and for his hospitality in Australia.

I would like to express my appreciation to Prof. Hisashi Okawa, Prof. Mitsuhiko Maesato, and Dr. Kazuya Otsubo at Kyoto University for their kind suggestions and beneficial discussions.

I am deeply grateful to Dr. Jared Taylor and Prof. Mohamedally Kurmoo at the University of Strasbourg, for their helpful suggestions on science and English writing, and for their kind support.

I offer my sincere thanks to Prof. Teppei Yamada at Kyushu University, Dr. Katsuyuki Morii at Nippon Shokubai Co., Ltd., Dr. Rie Makiura at Osaka Prefecture University, Prof. Katsuhiko Kanaizuka at Yamagata University, Prof. Lifen Yang at the Chinese Academy of Sciences, Prof. Yuki Nagao at the Japan Advanced Institute of Science and Technology, Dr. Akihito Shigematsu at Mitsui Chemicals, Inc., and Dr. Masaaki Sadakiyo at Kyushu University for their kind assistance and advice during my life as a researcher.

Special thanks go to all members in Prof. Hiroshi Kitagawa's laboratory for their many suggestions, their encouragement, and their friendship. Finally, I am thankful that I had the opportunity to spend time in this group for my research, and I will always be grateful to the people who helped me make it, especially my family.

Contents

Chapter 1
General Introduction

1.1 Alternative Energy

Stable and reliably supplied energy is required to maintain our comfort in daily life. Energy sources can be divided three broad categories; (1) chemical or photo physical energy through oxidizing reactions or absorbing sunlight, (2) nuclear energy, and (3) thermo mechanical energy in the form of wind or water and so on [1]. Each energy source has some disadvantages in terms of stability, cost or safety. While fossil and nuclear resources play the main role in global energy supply as it stands now, the alternative energy is most needed to keep our lives for the future because of the depletion of fossil fuel supplies and a public fear of nuclear accidents. Therefore many researchers all over the world are intensely studying the techniques to produce energy using solar cells [2], fuel cells [3], and to store energy utilizing hydrogen storage [4], rechargeable batteries [5] and high-temperature superconductivity [6] in order to replace of the current energy infrastructure. The most obvious source of alternative energy is the Sun. Solar cells make it an open possibility to capture the energy that is freely available from sunlight and convert it into valuable electric power [2]. Fuel cells directly provide electric energy from chemical energy with higher efficiency and lower emission of pollutants than conventional combustion [3]. Rechargeable batteries such as lithium-ion cells are more and more demanded as the portable energy by today's mobile society [5]. High-temperature superconductors have a potential to be the best materials for energy storage and energy transmission because there are no resistive losses [6]. Although these candidates have attractive advantages and they have been intensively investigated for years, each of them has still some difficulties which have to be overcome for practical use. Therefore they have not been able to play the main role yet.

K. Kusada, *Creation of New Metal Nanoparticles and Their Hydrogen-Storage and Catalytic Properties*, Springer Theses, DOI 10.1007/978-4-431-55087-7_1

1.2 The Issues of Hydrogen Utilization as an Alternative Energy Source

Hydrogen would be an ideal fuel because it is lightweight, highly abundant and its oxidation product (water) is environmentally benign [4]. To efficiently use the electric energy captured from unevenly-distributed natural energy including geothermal heat, tidal power and wind power, the development of highly efficient energy storage and conversion processes is required. Therefore, attention is focused on hydrogen as an electric energy-storage medium which enables the facile storage and transport of the electric energy [7]. However there are challenges to be overcome using this storage method. Conventional energy storage-methods are mainly high-pressure gas, liquid hydrogen or hydrocarbons, but these methods have some disadvantages. For example, it is difficult to treat high-pressure gas without over pressure and the compression itself is dangerous. Liquid hydrogen, of course, has a higher mass per unit volume than compressed gas, but the boiling temperature of normal hydrogen is 20 K [8]. Thus the direct loss of hydrogen occurs through heat transfer from the container. Hydrocarbons which are liquid at room temperature can be considered as a liquid hydrogen-storage medium, if they can be hydrogenated and dehydrogenated easily. Even though hydrogenation and dehydrogenation methods have been extensively studied, these methods performing under varying conditions have not been developed yet [4]. As for other storage methods, surface adsorption has attracted much attention. In particular, carbon species [9–11] and metal-organic frameworks (MOFs) [12–14] has been investigated as strong candidates because of their large surface area.

Figure 1.1 shows the storage capacities of carbonaceous adsorbents for hydrogen at 77 K and 1 bar. The amount of adsorbed hydrogen is proportional to the specific surface area of the carbon nanotube material and limited to 2 mass% for carbon materials [4]. MOFs are a class of porous hybrid materials built from metal ions and organic bridges and their structure and pore characteristics can

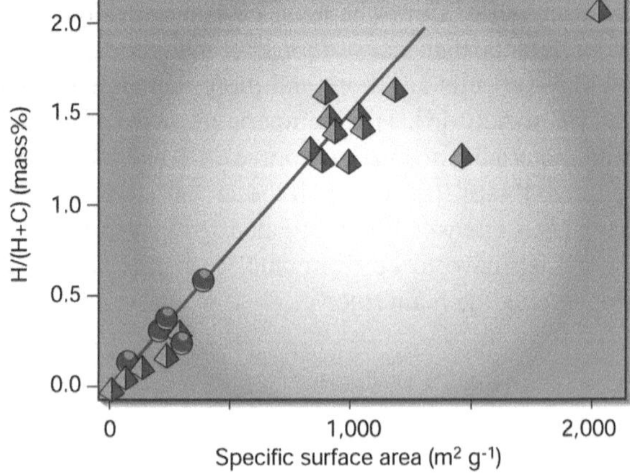

Fig. 1.1 Reversibly stored amount of hydrogen on various carbon materials versus the specific surface area of the samples. *Circles* represent nanotube samples (*best-fit line* indicated), *triangles* represent other nanostructured carbon samples [4]

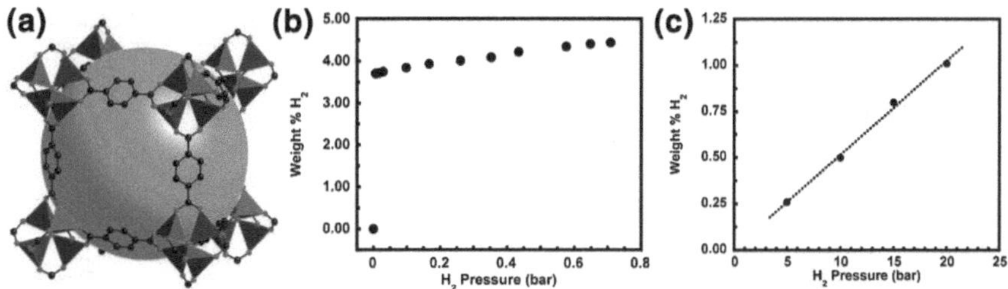

Fig. 1.2 **a** The single-crystal x-ray structure of MOF-5 illustrated for a single cube fragment of it cubic three-dimensional extended structure (Zn, *blue polyhedron*; O, *red sphere*; C, *black sphere*). The large *yellow sphere* represents the largest sphere that would fit in the cavities without touching the van der Waals atoms of the frameworks. Hydrogen atoms have been omitted. Hydrogen gas sorption isotherms for MOF-5 at **b** 78 K and **c** 298 K [12]

be tuned relative easily by changing metal ions or ligands. For example, MOF-5 (Zn_4O (BDC): BDC = 1,4-benzenedicarboxylate), one of the most famous MOFs, absorbs hydrogen up to 4.5 mass% at 78 K (Fig. 1.2) [12]. Although these high-surface-area materials show high total amount of adsorption per mass competing with other hydrogen storage materials, they are inconvenient to be used in our daily life because they need to be used under high pressure (about 50 bar), or low temperature (about 77 K). On the other hand, some metals and alloys have the capability to reversibly absorb large amount of hydrogen, and the metal hydride as a hydrogen storage material has much higher-safety than any of the other hydrogen storage candidates. Therefore metal hydrides are said to be the most promising materials for hydrogen storage [1].

But as metal is large element, their total amount of adsorption hydrogen per mass are lower than others (Fig. 1.3) and this problem needs to be solved for practical usage. A mechanism for the reaction of hydrogen with metal is as follows. Molecular hydrogen is dissociated into hydrogen atoms at the metal surface before absorption (Fig. 1.4a).

The some hydrogen atoms are initially dissolved in the host metal as a solid solution (α) phase. As the hydrogen pressure together with the absorption amount of hydrogen is increased, interaction between hydrogen atoms become locally important, and the nucleation and growth of hydride (β) phase start. While both of α and β phases exist together, the isotherms exhibits a flat plateau, the length of which represents how much hydrogen can be reversibly stored with small pressure change (Fig. 1.4b). Hydrogen is located in not the molecules but the atoms in the interstitial sites of the host metal lattice. During hydrogen absorption, the host metal lattice expands. In the pure β phase, the concentration of hydrogen in metal continuously increases with rising the hydrogen pressure. The two phase region ends at the critical point T_C, and the transition from α to β phase is continuous above this point [4].

In order to discover a more appropriate hydrogen storage metal for practical usage, basic scientific research is important. One of the most studied hydrogen

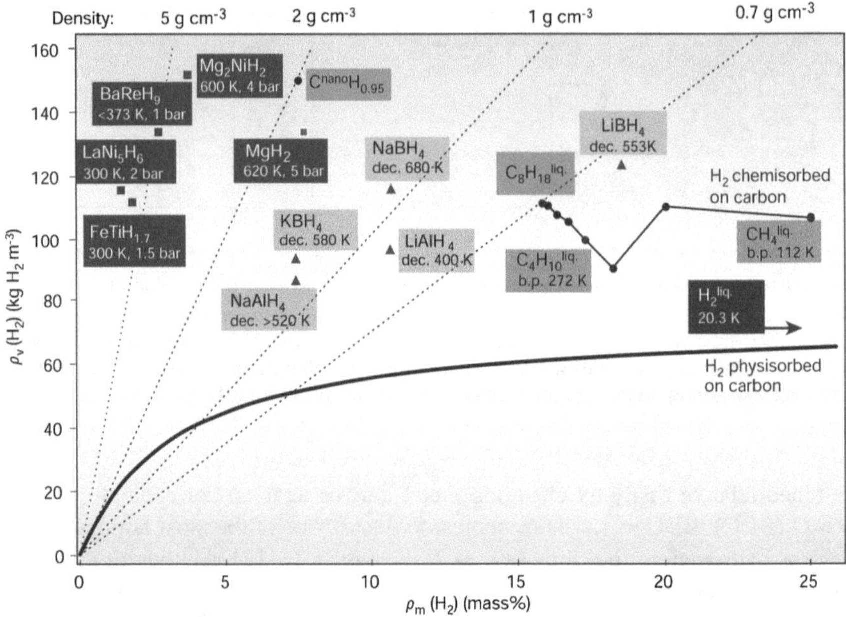

Fig. 1.3 Stored hydrogen per mass and per volume. Comparison of metal hydrides, carbon nanotubes and other hydrocarbons [4]

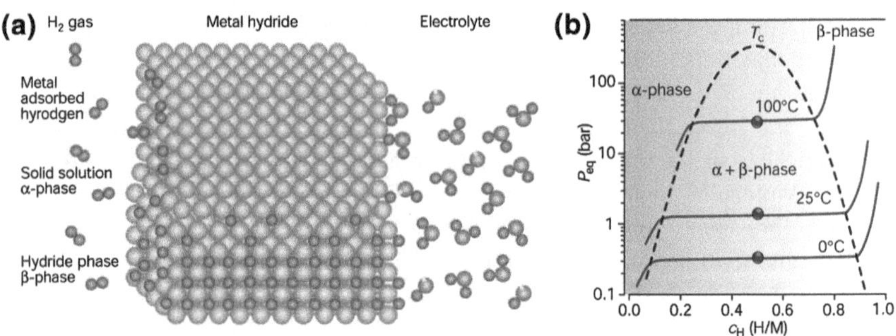

Fig. 1.4 **a** Schematic model of a metal structure with H atoms in the interstices between the metal atoms, and H_2 molecules at the surface. Hydrogen atoms are from physisorbed hydrogen molecules on the *left-hand side* and from the dissociation of water molecules on *right-hand side*. **b** Pressure-concentration-temperature plot; values are for $LANi_5$ [4]

storage metals is palladium (Pd), and it has been found that its hydrogen storage property is closely related with its electronic state [15, 16]. The band structure of Pd is shown Fig. 1.5. The characteristic of electronic states in transition metals is that the Fermi energy is located in the *d* band. As shown in Fig. 1.5, the Fermi energy of Pd where the total number of electrons becomes 10 is located in the 4*d*

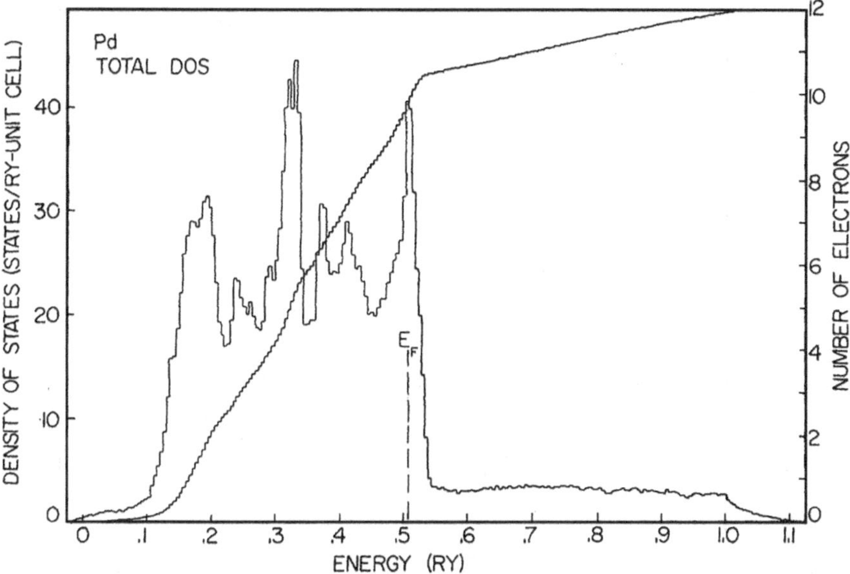

Fig. 1.5 Electronic density of states for Pd [17]

band, because the electron configuration of Pd is $5s^0 4d^{10}$ [17]. On the other hand, Fig. 1.6 shows the band structure of $PdH_{1.0}$. In comparison with Fig. 1.5, we can see the two new states. The one is generated below 4*d* band, and the other is generated at much higher level than the Fermi energy. The middle and bottom panels in Fig. 1.6 represent the calculated *s*-like density of state (DOS) for the H and Pd sites, respectively. It is observed that in the low energy region (ca. 0.1 Ry) the H site DOS is much higher than the Pd site DOS and a large fraction about 0.5 of the H states participate in the formation of the low-lying hydrogen(1 s)-palladium(4*d*) bonding states. Moreover from the top panel in Fig. 1.6, two electrons are contained in the lower energy states which are generated by the hydrogen absorption, and the total number of electrons becomes ten by the highest *d* peak located just below the Fermi energy. It should be noted that the DOS of pure Pd and of PdH above their high *d* peaks are quite similar and the total number of electrons below the highest *d* peak is the same for Pd and PdH (~10), regardless of the significant differences in the DOS well below the Fermi energy. Thus, the additional electron from the absorbed hydrogen raises the Fermi energy. From the calculation result, it was found that the Fermi energy for a hydrogen concentration of $PdH_{0.6}$ achieved the end of *d* band, [17] and it means that this concentration is enough to fill the holes in the *d* bands. Because the hydrogen capacity of Pd is 0.62 H/Pd under ambient conditions, the number of holes in the *d* band is related to the hydrogen capacity [18–20]. Another issue for practical usage of hydrogen as an alternative energy source is in the method of hydrogen production. While hydrogen is the most abundant chemical element in the universe, less than 1 % is exist as molecular

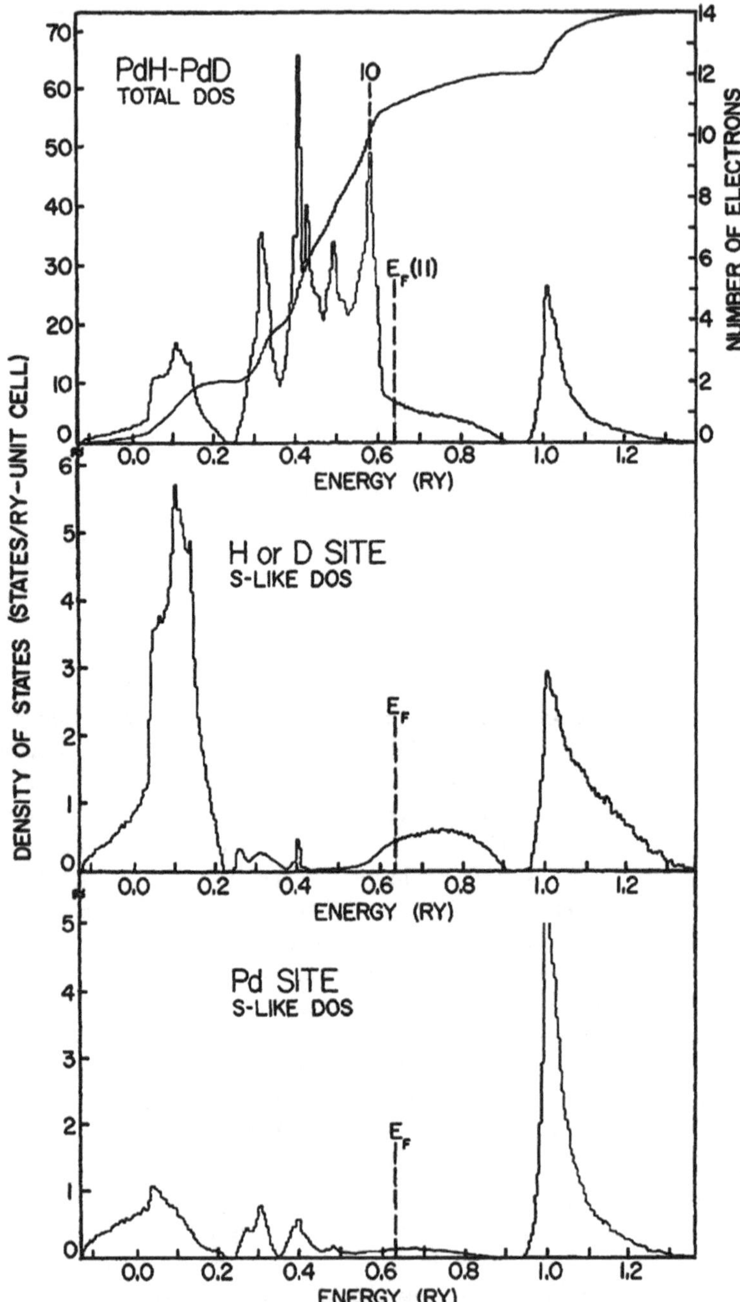

Fig. 1.6 Electronic density of states for $PdH_{1.0}$. *Top panel* shows the total DOS and *lower panel* shows the s-like DOS for H and Pd sites, respectively [17]

hydrogen gas. The overwhelming majority exists in water or hydrocarbons [4]. The clean way to produce hydrogen is, of course, the electrolysis of water as suggested by Grätzel [2]. However as it stands now, the most of the global hydrogen supply is obtained by reforming hydrocarbons [21, 22]. This 'reformed' hydrogen contains significant quantities carbon monoxide (CO) which poison hydrogen fuel cell devices. In addition because of the demand of CO removal from car exhaust, CO oxidation catalysts have been extensively investigated recently [23–30].

1.3 The Advantages in Solid Solution Alloy

An alloy is a "mixture" of more than one element, at least one of which is a metal element. The development of metallic alloys began in the Far East before 5,000 BC, but what we can call "scientific alloy research" started in the 19th century [31]. At that time, thermodynamics was applied to the phase equilibria of alloys, and the phase diagram was introduced as a basis of modern alloy science [32, 33]. Subsequently the systematic study of binary alloys was carried out by G. Tammann et al.

After that, conclusive evidence about the crystal structures of metals and alloys were found after the discovery of the X-ray [31]. The crystal structure of an alloy depends on the combination and chemical composition of constituent elements. There are several types of alloys; for example, "solid solution alloy" where the constituents are mixed at the atomic level, "eutectic alloy" where solid solution phases having different compositions coexist, and "phase-separated alloy" and so on. Because of the possibility of designing physical and chemical properties of a material by changing the combination and composition of constituents, alloys have been put in the forefront of materials from early human history up to our time. In particular a solid solution alloy is able to have continuously controlled chemical and physical properties by changing chemical compositions. Therefore, many types of solid solutions have been investigated for their solid-state properties including superconductivity, magnetism and others. As such for hydrogen storage materials, solid solution alloys have been eagerly studied. For instance Pd–Silver (Ag) and Pd–Rhodium (Rh) systems are representative examples which have been widely investigated for the reaction of hydrogen with metals [34–36]. As stated previously, Pd is a well-known hydrogen storage metal, and Rh, Pd and Ag are neighboring noble metals in the 4d transition-metal series. PdAg solid solution is easily obtained, since Ag and Pd can form solid solution having face-centered cubic (fcc) structure over the whole composition range as shown in Fig. 1.7a. In the electronic structure of PdAg, the number of holes in the *d* band is decreased with increasing Ag content (Fig. 1.7c). On the other hand as shown in Fig. 1.7b, Pd and Rh can mix each other at a temperature of above about 900 °C, but they cannot form solid solution below this temperature. As metallurgy is progressed and synthetic methods are developed, the solid solution is able to be obtained by using, for example, quenching technique. When Pd forms a solid solution with Rh,

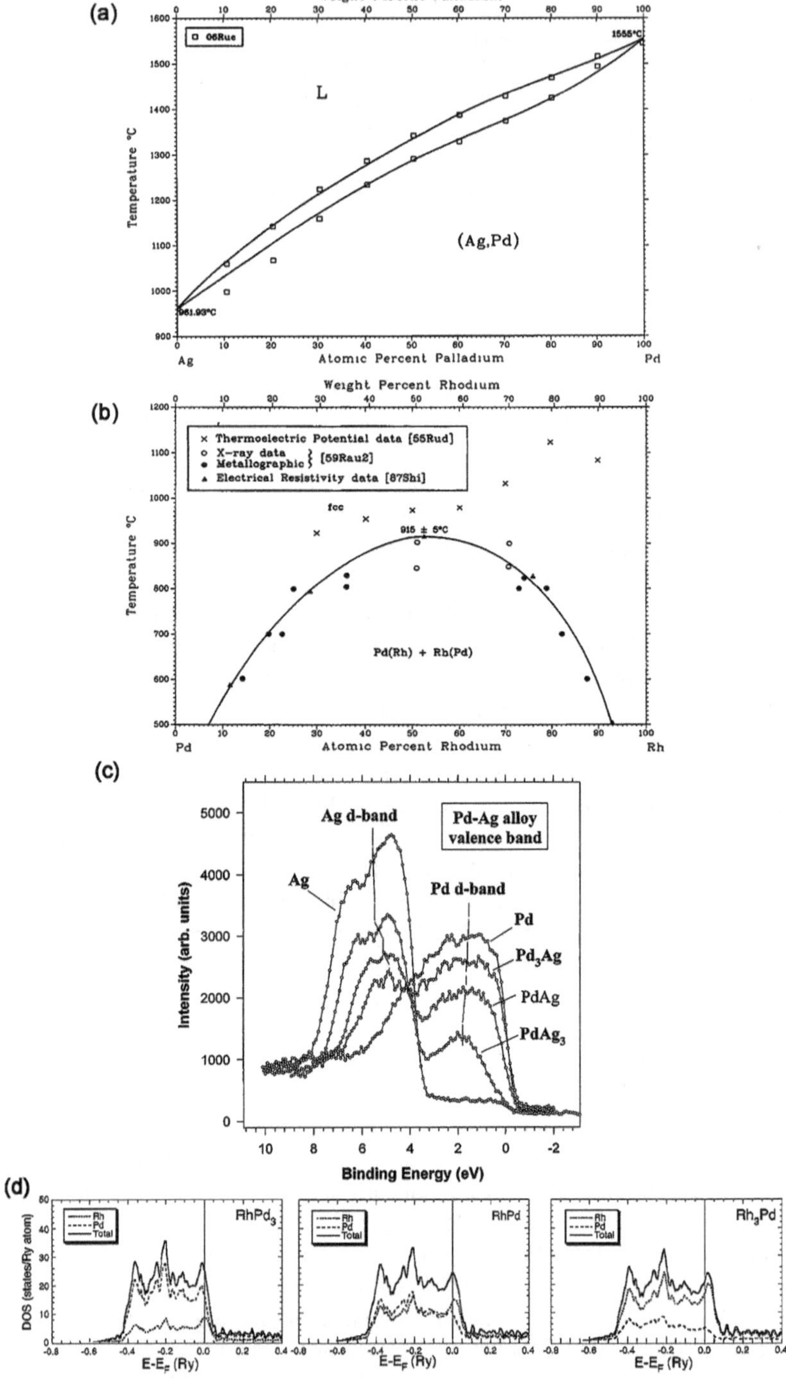

Fig. 1.7 The phase diagrams of **a** Pd–Ag [37] and (**b**) Pd–Rh systems [38]. **c** X-ray photoemission spectroscopy valence band spectra for PdAg alloys [39]. **d** Total and partial DOSs of PdRh alloys. The Fermi energy is indicated as the zero energy [40]

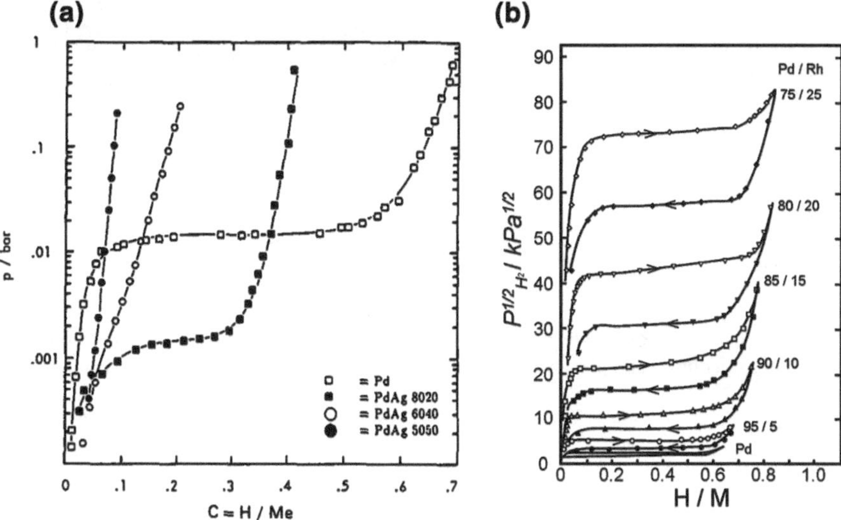

Fig. 1.8 The pressure-composition isotherms of **a** Pd_xAg_{100-x} [41] and **b** Pd_xRh_{100-x} alloys [36]

the change in the electronic structure of the alloy is different from Pd–Ag system (Fig. 1.7d). The number of holes in the *d* band of PdRh solid solution is increased with increasing Rh content. The hydrogen storage properties, including the capacity and the plateau pressure, in both of PdAg and PdRh solid solutions are continuously controlled by changing composition (Fig. 1.8). Therefore, since solid solution has the advantage of being able to continuously tune its properties, it has been used for creating new materials for the target properties.

1.4 The Possibility of Metal Nanoparticles as Advanced Materials

Recently, very small particles having size of from a few to several nanometers have attracted a lot of attention from both scientific and technological viewpoints, since these nanoparticles show unique properties different from those of the corresponding bulk materials.

The number of atoms in particles become few and the surface to volume ratio becomes large, as the size of nanoparticle decreases (Fig. 1.9) [42]. The one of the reasons for unique properties in nanoparticles is the "surface effect" where the surface atoms assume a larger role, because the surface atoms have some dangling-bonds and are in a higher-energy state than the inner atoms [43]. There is another effect on the properties of nanoparticle. The effect is "quantum size effect", where the continuous energy band in bulk material is changed into discrete band structure as the particle size decreases [44]. Anomalous phase behavior is well known as one

Full-Shell "Magic Number" Clusters					
Number of shells	1	2	3	4	5
Number of atoms in cluster	M_{13}	M_{55}	M_{147}	M_{309}	M_{561}
Percentage surface atoms	92%	76%	63%	52%	45%

Fig. 1.9 Idealized representation of fcc full-shell (Magic number) clusters. The numbers of shells, atoms in cluster and percentage surface atoms for each cluster are shown [42]

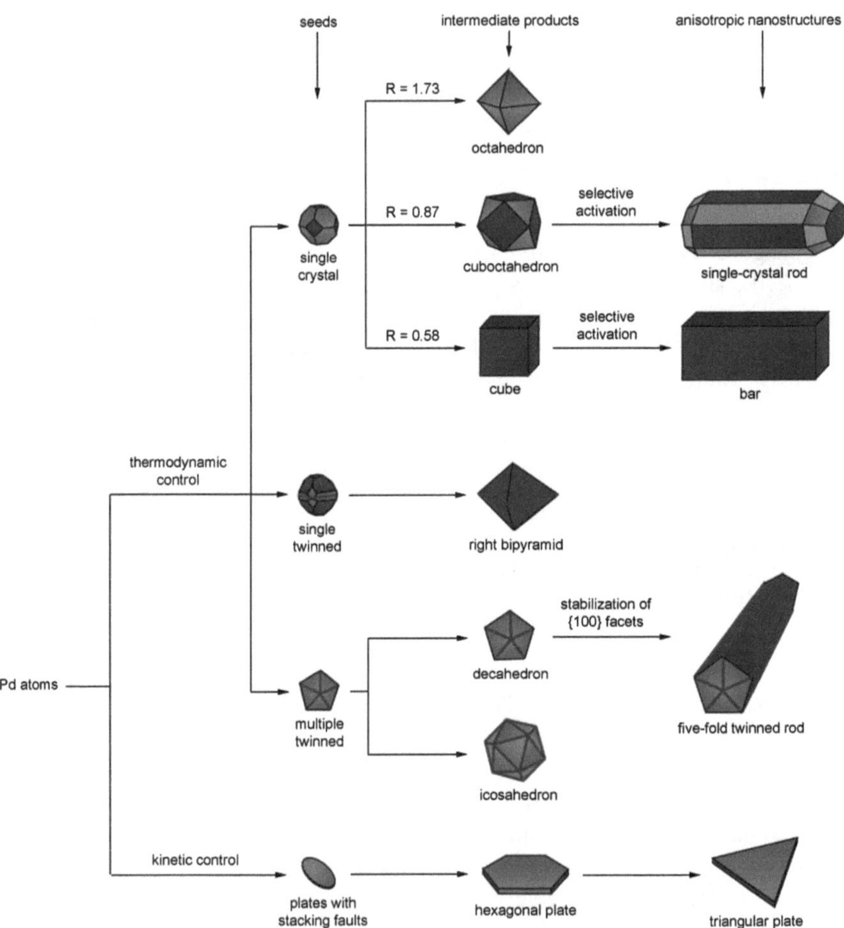

Fig. 1.10 A schematic of the reaction pathways that lead to Pd nanostructures with different shapes. The *green*, *orange* and *purple* colors represent the {100}, {111} and {110} facets, respectively [53]

of the unique properties in nanoparticles resulting from these effects. For example, phase transition temperatures such as the melting point are significantly changed with decreasing the particle size [45, 46]. Moreover, fcc Co and fcc Fe nanoparticles are stabilized at ambient conditions, even though these phases only exist at high temperature in bulk [47–49]. In chemical reactions, gold is one of the most famous examples exhibiting unique property due to nanosize-effect: although gold was earlier considered to be chemically inert and to have less catalytic activity, it turns out to be a highly active catalyst for many reactions such as CO oxidation when its size become less than 10 nm [50–52]. Metal nanoparticles can be synthesized by in a variety of methods in gas phase, in solution, supported on a substrate, or in a matrix [43]. The chemical reduction method in solution is the most intensively studied synthetic method. Nanoparticles are produced by chemical reduction of metal salts dissolved in a solvent with surfactant or polymeric ligands which passivate the cluster surface. Recently, using this method, not only the size but also the shape of particles is controlled to change the properties through controlling the growth rate of different crystallographic facets [53–56] (Fig. 1.10).

As with bulk metals, alloy nanoparticles also greatly extend the range of properties of metallic nanomaterials, and in recent years much attention has been paid to alloy nanoparticles as candidates for new functional materials. In alloy nanoparticles, the size of particles is a parameter of changing property in addition to the composition and combination of constituents. Figure 1.11 shows four main types of mixing patterns in bimetallic system. Core-shell segregated alloy (Fig. 1.11a) consists of a shell of one type of metal surrounding a core of another metal [58, 59]. Segregated alloys (Fig. 1.11c) consist of two kinds of metal subclusters [60]. Mixed alloys (Fig. 1.11b) can be separated into ordered (left) and random (solid solution, right) [61].

Multishell alloy (Fig. 1.11d) may present layered alternating shells [62]. Similar to the pure metal nanoparticles, anomalous phase behavior is observed

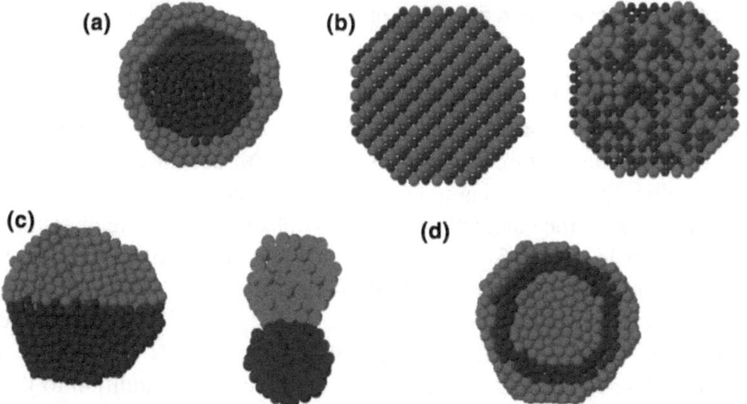

Fig. 1.11 A schematic of some possible mixing patterns: **a** core-shell, **b** solid solution, **c** subcluster segregated, **d** three shell. The pictures show cross sections of the nanoparticles [57]

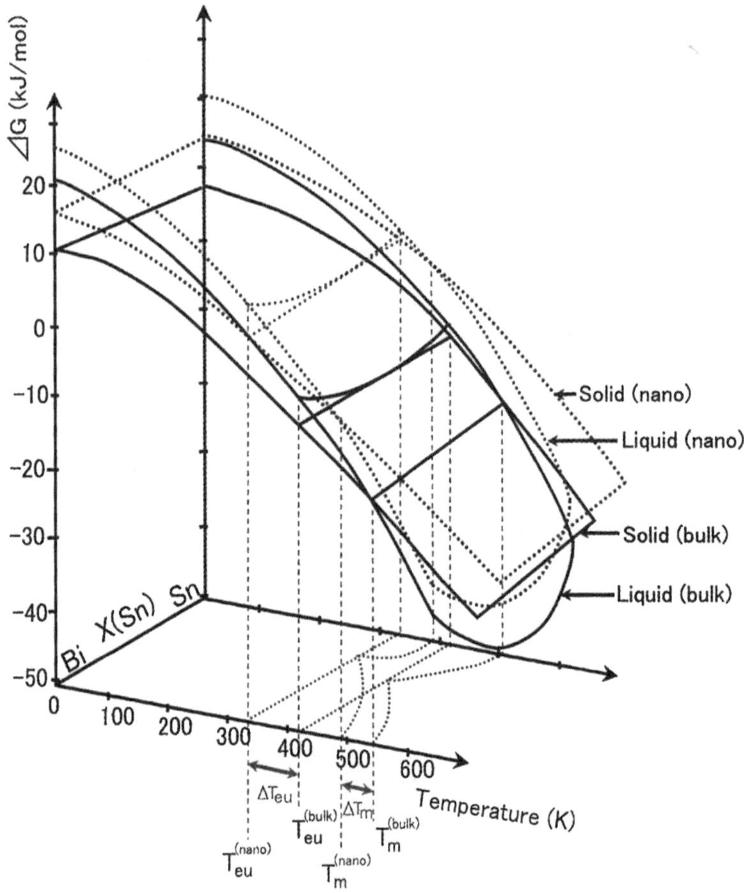

Fig. 1.12 Gibbs free energy-temperature—composition diagram in the Bi-Sn system. The *solid* and *dotted lines* indicate the Gibbs free energy of bulk and 10 nm particles, respectively [31]

in alloy nanoparticles. For example, enhanced atom mixing such as spontaneous alloying at room temperature in Au-Ag and Au-Cu systems were observed [63, 64]. The eutectic temperature (T_{eu}) is also lowers with decreasing the particle size similar to melting-point lowering in pure metal nanoparticles.

Figure 1.12 shows the calculation result of Gibbs free energy-temperature-composition diagram in the Bi-Sn system [31]. In order to quantify the phase equilibrium of nanoparticles, not only the Gibbs free energy in the bulk (such as the volume free energy), but also the surface free energy has to be considered because of the extremely large surface-to-volume ratio in nanoparticles. Figure 1.12, represents the decrease of T_{eu} with decreasing the particle size in addition to the melting-point lowering. Furthermore, the enhancement of the solid solubility in Pb-Sn alloy nanoparticles has been observed [65]. Although only 10 at.% Sn can mix in lead at 383 K in bulk, 17 nm alloy nanoparticles contained 56 at.% Sn at the same temperature, which is about five times higher than the solubility limit of tin in bulk lead [65]. Moreover, recently there is a great development in many synthesis methods easily producing the solid solution nanoparticles, such as AuPt, AuNi, AuFe, AgPt solid solutions, which can only be obtained in a metastable state in bulk systems [66–72]. In the nanometer scale, there is a possibility of creating advanced materials which cannot be obtained in bulk.

1.5 The Structure of This Thesis

The purpose of this thesis is to report the first discovery of metal nanoparticles having new structure and exhibiting hydrogen storage ability or CO oxidation activity. This section gives a short overview of the chapters and their interconnections. Chapter 2, "Hydrogen-Storage Properties of Solid Solution Alloys of Immiscible Neighboring Elements with Pd," is dedicated to the example of a AgRh solid solution alloy exhibiting hydrogen-storage properties. Rh and Ag are the neighboring elements with Pd, which is a well-known hydrogen storage metal. However, Rh and Ag do not possess hydrogen storage properties. AgRh solid solution alloys have not explored in the past because they do not mix each other at the atomic level even in the liquid phase. In Chap. 3, "Systematic study of the Hydrogen Storage Properties and the CO-oxidizing Abilities of Solid Solution Alloy Nanoparticles in an Immiscible Pd–Ru System," it is shown that the compositional dependence of the hydrogen storage properties and CO-oxidizing ability in PdRu nanoparticles. While Pd has face-centered cubic (fcc) structure, Ru has hexagonal close packing (hcp) structure. Thus the compositional dependence of alloy's structure is another topic in this chapter. The first discovery of pure fcc Ru nanoparticles is given in Chap. 4, "Discovery of the Face-centered Cubic Ruthenium Nanoparticles: Facile Size-controlled Synthesis using the Chemical Reduction Method". In this chapter, it is also reported that the CO-oxidizing activity over Ru nanoparticles depends on their structure and size. Chapter 5, "Changeover of the Thermodynamic Behavior for Hydrogen Storage in Rh with Increasing Nanoparticle Size," represents the size dependence of hydrogen storage properties in Rh nanoparticles. The enthalpy of hydrogen storage in Rh nanoparticles was changed from exothermic to endothermic with increasing particle size, with a critical size between 7.1 and 10.5 nm.

References

1. Dresselhaus MS, Thomas IL (2001) Alternative energy technologies. Nature 414:332–337
2. Grätzel M (2001) Photoelectrochemical cells. Nature 414:338–344
3. Steele BCH, Heinzel A (2001) Materials for fuel-cell technologies. Nature 414:345–352
4. Schlapbach L, Züttel A (2001) Hydrogen-storage materials for mobile applications. Nature 414:353–358

5. Tarascon JM, Armand M (2001) Issues and challenges facing rechargeable lithium batteries. Nature 414:359–367
6. Larbalestier D, Gurevich A, Feldmann DM, Polyanskii A (2001) High-Tc superconducting materials for electric power applications. Nature 414:368–377
7. Tamura H (1998) Hydrogen storage alloys—fundamentals and frontier technologies. NTS Inc., p 30
8. Ancsin J (1977) Thermometric fixed points of hydrogen. Metrologia 13:79–86
9. Dillon AC, Jones KM, Bekkedahl TA, Kiang CH, Bethune DS, Heben MJ (1997) Storage of hydrogen in single-walled carbon. Nature 386:377–379
10. Baughman RH, Zakhidov AA, de Heer WA (2002) Carbon nanotubes—the route toward applications. Science 297:787–792
11. Jiménez V, Ramírez-Lucas A, Sánchez P, Valverde JL, Romero A (2012) Improving hydrogen storage in modified carbon materials. Int J Hydrogen Energy 37:4144–4160
12. Rosi NL, Eckert J, Eddaoudi M, Vodak DT, Kim J, O'Keeffe M, Yaghi OM (2003) Hydrogen storage in microporous metal-organic frameworks. Science 300:1127–1129
13. Farha OK, Yazaydin Aö, Eryazici I, Malliakas CD, Hauser BG, Kanatzidis MG, Nguyen ST, Snurr RQ, Hupp JT (2010) De novo synthesis of a metal–organic framework material featuring ultrahigh surface area and gas storage capacities. Nat Chem 2:944–948
14. Ward MD (2003) Molecular Fuel Tanks. Science 300:1104–1105
15. Alefeld G, Völkl J (1978) Hydrogen in metals I. Springer, Berlin, Heidelberg, New York
16. Alefeld G, Völkl J (1978) Hydrogen in metals II. Springer, Berlin, Heidelberg, New York
17. Papaconstantopoulos DA, Klein BM, Economou EN, Boyer LL (1978) Band structure and superconductivity of PdD_X and PdH_X. Phys Rev B 17:141–150
18. Vuillemin JJ, Priestly MG (1965) De Haas-Van Alphen effect and fermi surface in palladium. Phys Rev Lett 14:307–308
19. Mueller FM, Freeman AJ, Dimmock JO, Furdyna AM (1970) Electronic structure of palladium. Phys Rev B 1:4617–4634
20. Wicke E (1984) Electronic structure and properties of hydrides of 3d and 4d metals and intermetallics. J Less-Common Met 101:17–33
21. Besenbacher F, Chorkendorff I, Clausen BS, Hammer B, Molenbroek AM, Nørskov JK, Stensgaard I (1998) Design of a surface alloy catalyst for steam reforming. Science 279:1913–1915
22. Huber GW, Shabaker JW, Dumesic JA (2003) Raney Ni-Sn catalyst for H_2 production from biomass-derived hydrocarbons. Science 300:2075–2077
23. Ertl G (2008) Reactions at surfaces: from atoms to complexity (Nobel lecture). Angew Chem Int Ed 47:3524–3535
24. Xie X, Li Y, Liu Z, Haruta M, Shen W (2009) Low-temperature oxidation of CO catalysed by Co_3O_4 nanorods. Nature 458:746–749
25. Kaden WE, Wu T, Kunkel WA, Anderson SL (2009) Electronic structure controls reactivity of size-selected Pd clusters adsorbed on TiO_2 surfaces. Science 326:826–829
26. Alayoglu S, Nilekar AU, Mavrikakis M, Eichhorn B (2008) Ru–Pt core–shell nanoparticles for preferential oxidation of carbon monoxide in hydrogen. Nat Mater 7:333–338
27. Qiao B, Wang A, Yang X, Allard LF, Jiang Z, Cui Y, Liu J, Li J, Zhang T (2011) Single-atom catalysis of COoxidation using Pt_1/FeO_x. Nat Chem 3:634–641
28. Roth C, Benker N, Buhrmester T, Mazurek M, Loster M, Fuess H, Koningsberger DC, Ramaker DE (2005) Determination of O[H] and CO coverage and adsorption sites on PtRu electrodes in an operating PEM fuel cell. J Am Chem Soc 127:14607–14615
29. Jiang H, Liu B, Akita T, Haruta M, Sakurai H, Xu Q (2009) Au@ZIF-8: CO oxidation over gold nanoparticles deposited to metal-organic framework. J Am Chem Soc 131:11302–11303
30. Kim HY, Lee HM, Henkelman G (2012) CO oxidation mechanism on CeO_2-supported Au nanoparticles. J Am Chem Soc 134:1560–1570
31. Pfeiler W (2007) Alloy physics. Wiley-VCH, New Jersey
32. Gibbs JW (1976) The equilibrium of heterogeneous substances. Trans Conn Acad 3:108–248

33. Gibbs JW (1978) The equilibrium of heterogeneous substances (concluded). Trans Conn Acad 3:343–524
34. Fazle Kibria AKM, Sakamoto Y (2000) The effect of alloying of palladium with silver and rhodium on the hydrogen solubility, miscibility gap and hysteresis. Int J Hydrogen Energy 25:853–859
35. Zeng G, Goldbach A, Shi L, Xu H (2012) Compensation effect in H_2 permeation kinetics of PdAg membranes. J Phys Chem C 116:18107
36. Noh H, Luo W, Flanagan TB (1993) The effect of annealing pretreatment of Pd-Rh alloys on their hydrogen solubilities and thermodynamic parameters for H_2 solution. J Alloys Compd 196:7–16
37. Karakaya I, Thompson WT (1988) The Ag-Pd (silver-palladium) system. Bull Alloy Phase Diagrams 9:237–243
38. Tripathi SN, Bharadwaj SR (1944) The Pd-Rh (palladium-rhodium) system. J Phase Equil 15:208–212
39. Coulthard I, Sham TK (1996) Charge redistribution in Pd-Ag alloys from a local perspective. Phys Rev Lett 77:4824–4827
40. Turchi PEA, Drchal V, Kudrnovský J (2006) Stability and ordering properties of fcc alloys based on Rh, Ir, Pd, and Pt. Phys Rev B 74:064202–064212
41. Züchner H, Rauf T (1991) Electrochemical isotherm measurements on the Pd-H and PdAg-H systems. J Less-Common Met 172–174:816–823
42. Aiken JD III, Finke RG (1999) A review of modern transition-metal nanoclusters: their synthesis, characterization, and applications in catalysis. J Mol Catal A Chem 145:1–44
43. Owens FJ, Poole CP Jr (2008) The physics and chemistry of nanosolids. Wiley Interscience, New Jersey
44. Kubo R (1962) Electronic properties of metallic fine particles I. J Phys Soc Jpn 17:975–986
45. Zhang M, Efremov MY, Schiettekatte F, Olson EA, Kwan AT, Lai SL, Wisleder T, Greene JE, Allen LH (2000) Size-dependent melting point depression of nanostructures: Nanocalorimetric measurements. Phys Rev B 62:10548–10564
46. Blackman M, Sambles JR (1970) Melting of very small particles during evaporation at constant temperature. Nature 226:938
47. Dong XL, Choi CJ, Kim BK (2002) Chemical synthesis of Co nanoparticles by chemical vapor condensation. Scr Mater 47:857–861
48. Ling T, Xie L, Zhu J, Yu H, Ye H, Yu R, Cheng Z, Liu L, Yang G, Cheng Z, Wang Y, Ma X (2009) Icosahedral face-centered cubic Fe nanoparticles: facile synthesis and characterization with aberration-corrected TEM. Nano Lett 9:1572–1576
49. Kim H, Kaufman MJ, Sigmund WM, Jacques D, Andrews R (2003) Observation and formation mechanism of stable face-centered-cubic Fe nanorods in carbon nanotubes. J Mater Res 18:1104–1108
50. Ishida T, Haruta M (2007) Gold catalysts: towards sustainable chemistry. Angew Chem Int Ed 46:7154–7156
51. Haruta M, Tsubota S, Kobayashi T, Kageyama H, Gent MJ, Delmon B (1993) Low-temperature oxidation of CO over gold supported on TiO_2, α-Fe_2O_3, and Co_3O_4. J Catal 144:175–192
52. Haruta M, Yamada N, Kobayashi T, Iijima S (1989) Gold catalysts prepared by coprecipitation for low-temperature oxidation of hydrogen and of carbon monoxide. J Catal 115:301–309
53. Xiong Y, Xia Y (2007) Shape-controlled synthesis of metal nanostructures: the case of palladium. Adv Mater 19:3385–3391
54. Lim B, Jiang M, Tao J, Camargo PHC, Zhu Y, Xia Y (2009) Shape-controlled synthesis of Pd nanocrystals in aqueous solutions. Adv Funct Mater 19:189–200
55. Lee I, Delbecq F, Morales R, Albiter MA, Zaera F (2009) Tuning selectivity in catalysis by controlling particle shape. Nat Mater 8:132–138
56. Tao A, Sinsermsuksakul P, Yang P (2007) Tunable plasmenic lattices of silver nanocrystals. Nat Nanotechnol 2:435–440

57. Ferrando R, Jellinek J, Johnston RL (2008) Nanoalloys: from theory to applications of alloy clusters and nanoparticles. Chem Rev 108:845–910
58. Tedsree K, Li T, Jones S, Chan CWA, Yu KMK, Bagot PAJ, Marquis EA, Smith GDE, Tsang SCE (2011) Hydrogen production from formic acid decomposition at room temperature using a Ag–Pd core—shell nanocatalyst. Nat Nanotechnol 6:302–307
59. Serpell CJ, Cookson J, Ozkaya D, Beer PD (2011) Core@ shell bimetallic nanoparticle synthesis via anion coordination. Nat Chem 3:478–483
60. Saruyama M, So Y, Kimoto K, Taguchi S, Kanemitsu Y, Teranishi T (2011) Spontaneous formation of Wurzite-CdS/Zinc blende-CdTe heterodimers through a partial anion exchange reaction. J Am Chem Soc 133:17598–17601
61. Shun TT, Hung CH, Lee CF (2010) Formation of ordered/disordered nanoparticles in FCC high entropy alloys. J Alloys Compd 493:105–109
62. Anh DTN, Singh P, Shankar C, Mott D, Maenosono S (2011) Charge-transfer-induced suppression of galvanic replacement and synthesis of (Au@Ag)@Au double shell nanoparticles for highly uniform, robust and sensitive bioprobes. Appl Phys Lett 99:073107
63. Shibata T, Bunker BA, Zhang Z, Meisel D, Vardeman CF, Gezelter JD (2002) Size-dependent spontaneous alloying of Au-Ag nanoparticles. J Am Chem Soc 124:11989–11996
64. Yasuda H, Mori H (1994) Cluster-size dependence of alloying behavior in gold clusters. Z Phys D 31:131–134
65. Lee JG, Mori H (2004) Direct evidence for reversible diffusional phase change in nanometer-sized alloy particle. Phys Rev Lett 93:235501–235504
66. Zhou S, Jackson GS, Eichhorn B (2007) AuPt alloy nanoparticles for CO-tolerant hydrogen activation: architectural effects in Au-Pt bimetallic nanocatalysts. Adv Funct Mater 17:3099–3104
67. Hernández-Fernández P, Rojas S, Ocón P, Gómez de la Fuente JL, San Fabián J, Sanza J, Peña MA, García-García FJ, Terreros P, Fierro JLG (2007) Influence of the preparation route of bimetallic Pt–Au nanoparticle electrocatalysts for the oxygen reduction reaction. J Phys Chem C 111:2913–2923
68. Lang H, Maldonado S, Stevenson KJ, Chandler BD (2004) Synthesis and characterization of dendrimer templated supported bimetallic Pt–Au nanoparticles. J Am Chem Soc 126:12949–12956
69. Chiang I, Chen Y, Chen D (2009) Synthesis of NiAu colloidal nanocrystals with kinetically tunable properties. J Alloys Compd 468:237–245
70. Lu D, Domen K, Tanaka K (2002) Electrodeposited Au–Fe, Au–Ni, and Au–Co alloy nanoparticles from aqueous electrolytes. Langmuir 18:3226–3232
71. Chiang I, Chen D (2007) Synthesis of monodisperse FeAu nanoparticles with tunable magnetic and optical properties. Adv Funct Mater 17:1311–1316
72. Torigoe K, Nakajima Y, Esumi K (1993) Preparation and characterization of colloidal silver-platinum alloys. J Phys Chem 97:8304–8309

Chapter 2
Hydrogen Storage Properties of Solid Solution Alloys of Immiscible Neighboring Elements with Pd

2.1 Introduction

Atomic-level (solid solution) alloying has the advantage of being able to continuously control chemical and physical properties of elements by changing compositions and/or combinations of constituent elements [1–3]. However, solid solution phases in alloys are limited to specific combinations of elements. Furthermore, most of the combinations have solid solution phases in limited regions of composition and temperature. Thus, alloys often undergo phase separation from the high-temperature solid solution state when the temperature is reduced. Even in such cases, we can obtain solid solution phases at room temperature as a metastable state using a quenching technique from the high-temperature solid solution phase. However, in certain cases, the constituent elements are immiscible even in a high-temperature liquid phase. In such cases, solid solution phases have not yet been obtained, because the quenching technique is not applicable to the phase-separated liquid phase.

Rh, Pd and Ag are neighboring noble metals in the 4d transition-metal series. Each is well known as an effective catalyst for various chemical reactions [4–7]. If these metals could be mixed in a desired ratio, then their chemical and physical properties could be enhanced. In the Ag–Pd solid solution system that has complete solid solubility, the hydrogen permeability is most enhanced when the ratio is Ag:Pd = 24:76. The thin film resulting from this system is used as a hydrogen-permeable membrane [8]. However, in the other two combinations, namely Pd–Rh and Ag–Rh, phase separation occurs, with complete insolubility at room temperature [9, 10]. In the case of Ag–Rh, even in the liquid phase at around 2,000 °C, Ag and Rh do not mix, and segregated clustering of each element forms. Considering the band-filling effect in the III–V semiconductors, [11] the $Ag_{0.5}Rh_{0.5}$ solid solution alloy is expected to have a similar electronic structure to Pd and resemble Pd in terms of chemical and physical properties, since Pd is located between Rh and Ag in the periodic table. For example, one of the well-known unique properties of

K. Kusada, *Creation of New Metal Nanoparticles and Their Hydrogen-Storage and Catalytic Properties*, Springer Theses, DOI 10.1007/978-4-431-55087-7_2

Pd is its hydrogen storage property, which is attributed to its electronic structure; [12–15] Rh and Ag have no hydrogen storage ability. Does the $Ag_{0.5}Rh_{0.5}$ solid solution alloy actually absorb hydrogen? In this chapter, the author reports the first example of a AgRh solid solution alloy exhibiting a hydrogen storage property.

Recently, several techniques for stabilizing nonequilibrium phase at ordinary temperatures and pressures have been attracting attention. Changing the size of a particle to the nanometer scale appears to be a particularly efficient technique. For example, various nonequilibrium solid solution alloys such as Au–Pt, Au–Ni, Au–Fe and Ag–Pt, which exist in solid solution at high temperature or in the liquid phase, have been easily obtained as metal nanoparticles by solution chemistry methods [16–23]. This may allow the isolation of novel solid solution alloys even in the entirely immiscible Ag–Rh system.

2.2 Experiment

2.2.1 Syntheses of Ag_xRh_{1-x} Nanoparticles

In a typical synthesis of AgRh nanoparticles, poly(*N*-vinyl-2-pyrrolidone) (PVP, 111 mg, MW $\approx$ 40,000, Wako) was dissolved in ethylene glycol (200 ml) and the solution heated to 170 °C in air under magnetic stirring. Meanwhile, $AgNO_3$ (8.4 mg, 0.050 mmol, Kishida) and $Rh(CH_3COO)_3$ (14.1 mg, 0.050 mmol, Mitsuwa) were co-dissolved in deionized water (20 ml). The aqueous solution of $AgNO_3$ and $Rh(CH_3COO)_3$ was then added to the ethylene glycol. Other Ag_xRh_{1-x} ($x = 0.4$ and 0.7) nanoparticles were prepared by controlling the molar ratio of Ag^+ and Rh^{3+} ions. For $Ag_{0.4}Rh_{0.6}$ nanoparticles, $AgNO_3$ (4.2 mg, 0.025 mmol) and $Rh(CH_3COO)_3$ (20.9 mg, 0.075 mmol) were co-dissolved in deionized water (20 ml). For $Ag_{0.7}Rh_{0.3}$ nanoparticles, $AgNO_3$ (12.8 mg, 0.075 mmol) and $Rh(CH_3COO)_3$ (7.0 mg, 0.025 mmol) were co-dissolved in deionized water (20 ml).

2.2.2 Characterizations

Transmission Electron Microscopy Analysis

Transmission Electron Microscopy (TEM) images were recorded on a JEOL JEM-2000EX TEM instrument operated at 200 kV accelerate voltage. The samples dispersed with ethanol were drop-cast onto a carbon-coated copper grid and allowed to dry under ambient conditions.

Synchrotron X-Ray Diffraction Measurement

Synchrotron XRD patterns were measured at the beam line BL02B2, SPring-8. The diffraction patterns of the samples sealed in glass capillaries under vacuum were measured at 303 K. The wavelength is $\lambda = 0.55277$ Å.

Atomic Absorption Spectrophotometric Measurement

AAS analysis was carried out with a SHIMADZU AA-6300 using samples dissolved in nitric acid solution.

Scanning Transmission Electron Microscopy Analysis

High resolution (HR) STEM, high-angle annular dark-field (HAADF) STEM and energy-dispersive X-ray (EDX) analyses were recorded on a HITACHI HD-2700 STEM operated at 200 kV accelerate voltage at the Naka Application Center, Hitachi High-Technologies Corporation.

UV-Vis Spectra Measurement

UV-vis absorption spectra were measured for ethanol suspensions of Ag_xRh_{1-x} nanoparticles using a Jasco V-570 spectrophotometer.

In Situ XRD Measurement

The thermal stability of $Ag_{0.5}Rh_{0.5}$ nanoparticles was investigated by in situ powder XRD analysis measured at the BL02B2 beamline, SPring-8. The XRD patterns of the samples sealed in a glass capillary were measured in situ under vacuum condition at temperature range between 303 K and 573 K. The wavelength is 0.578375(1) Å.

Pressure-Composition Isotherms Measurement

PC isotherms were measured at hydrogen pressure range between 0 and ca. 100 kPa at 303 K by a volumetric technique using a pressure-composition-temperature apparatus (Suzuki Shokan Co., Ltd).

Solid-State ^2H NMR Spectra Measurement

Solid-state ^2H NMR spectra were recorded at 303 K in a fixed magnetic field of 94 kOe and a frequency sweep of 60.94–61.94 MHz, using a BRUKER NMR spectrometer. Each sample was evacuated in a glass capillary for several hours at 373 K, and then each sample was sealed into a glass capillary with 86.7 kPa of ^2H gas at 303 K.

X-ray Photoelectron Spectroscopy Analysis

XPS spectra for samples on an carbon sheet were recorded on a Physical Electronics PHI 5800 ESCA system with a background pressure of approximately 1×10^{-9} Torr.

2.3 Results and Discussion

TEM images of the synthesized Ag_xRh_{1-x} nanoparticles were recorded on a JEOL JEM-2000EX TEM instrument (Fig. 2.1). The mean diameters of the nanoparticles were determined from the TEM images to be (a) 10.2 ± 1.0, (b) 12.2 ± 1.6, and (c) 13.6 ± 1.5, respectively. Numbers that follow the ± sign represent estimated standard deviations. The size distributions were narrow and the mean diameters were estimated by averaging over at least 300 particles.

The crystal structures of Ag_xRh_{1-x} bimetallic nanoparticles were investigated by synchrotron X-ray ($\lambda = 0.55277$ Å) powder diffraction at the beam line BL02B2, SPring-8. The diffraction patterns of the samples sealed in glass capillaries under vacuum were measured at 303 K. Figure 2.2 shows the XRD patterns of the Ag and Rh, and three compositions of AgRh nanoparticles ($Ag_{0.7}Rh_{0.3}$, $Ag_{0.5}Rh_{0.5}$ and $Ag_{0.4}Rh_{0.6}$). All the AgRh samples showed diffraction patterns consistent with single face-centered-cubic structure without signals from pure

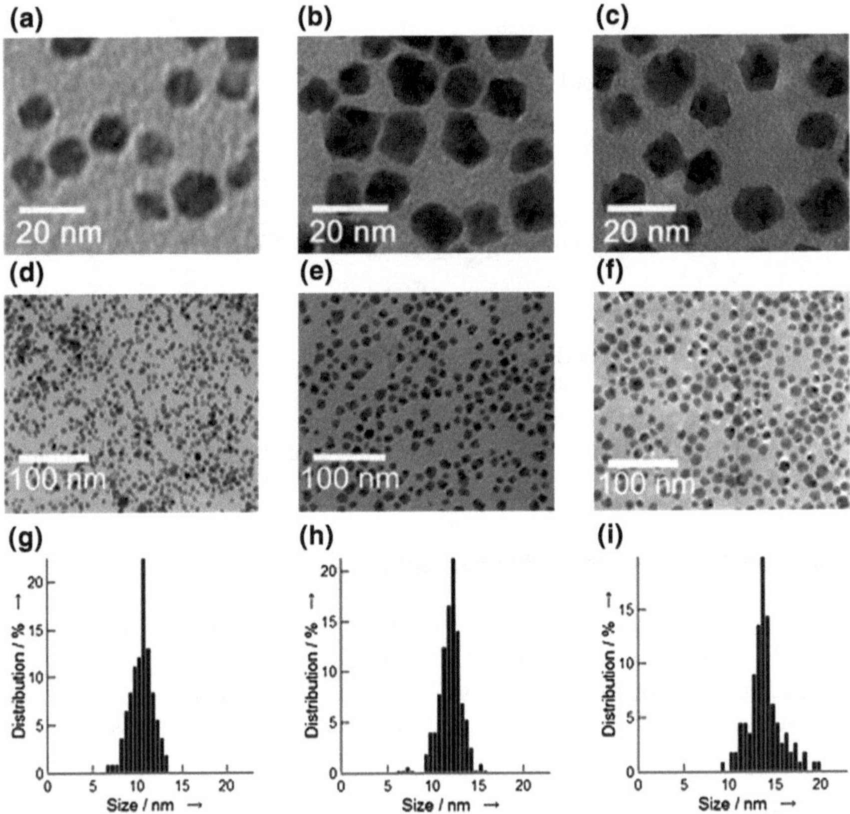

Fig. 2.1 The TEM images and the size distributions of **a, d, g** $Ag_{0.4}Rh_{0.6}$, **b, e, h** $Ag_{0.5}Rh_{0.5}$, and **c, f, i** $Ag_{0.7}Rh_{0.3}$

Ag or Rh phases (Fig. 2.2a). The diffraction peaks of AgRh nanoparticles shifted continuously to the higher-angle side with increasing Rh content in the AgRh nanoparticles, which is in agreement with the shrinkage of the lattice parameter due to the smaller unit cell parameter of Rh (Fig. 2.2b). This result strongly supports the formation of the atomic-level AgRh alloy.

To investigate the composition of Ag and Rh atoms in the nanoparticles, elemental analyses were carried out using AAS and EDX techniques (Table 2.1). The average stoichiometry from the AAS data was Ag:Rh = 0.51:0.49 and that from the EDX data was Ag:Rh = 0.49:0.51.

Figure 2.3 shows elemental mapping data for a group of prepared $Ag_{0.5}Rh_{0.5}$ nanoparticles. Figure 2.3a shows a HAADF-STEM image. Figure 2.3b and c show the corresponding Ag–L and Rh–L STEM-EDX maps, respectively. Figure 2.3d presents an overlay map of the Ag and Rh chemical distribution. The mapping data (Fig. 2.3d) provides visually apparent evidence of the formation of AgRh solid solutions.

The author further characterized the AgRh nanoparticles by EDX line scanning analysis (Fig. 2.4). The line-scan position of the nanoparticle is denoted by the white line in the inset of Fig. 2.4. The compositional line profiles of Ag and Rh on a $Ag_{0.5}Rh_{0.5}$ nanoparticle show that atomic-level Ag–Rh alloying successfully occurs (Fig. 2.5).

In order to investigate the change of the optical properties accompanied by the formation of AgRh solid solution, UV-Vis absorption spectra were measured with

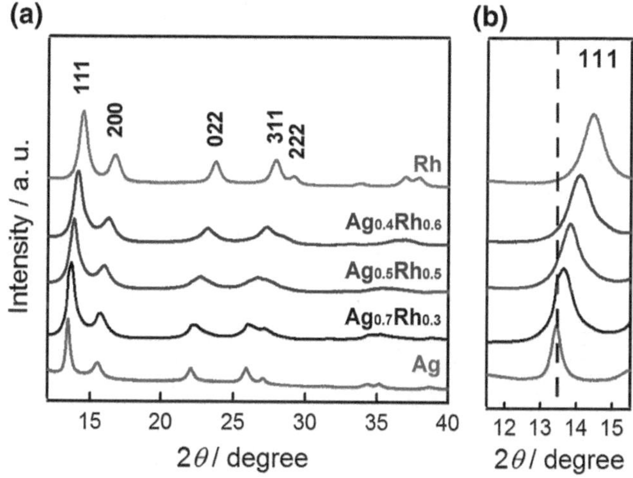

Fig. 2.2 **a** Synchrotron X-ray powder diffraction patterns ($2\theta = 12°–40°$) of Ag, Rh, and Ag–Rh nanoparticles at 303 K, **b** ($2\theta = 11°–16°$)

Table 2.1 Results of the particle composition of Ag–Rh bimetallic nanoparticles

AAS results	EDX results
$Ag_{0.4}Rh_{0.6}$	$Ag_{0.38}Rh_{0.62}$
$Ag_{0.51}Rh_{0.49}$	$Ag_{0.49}Rh_{0.51}$
$Ag_{0.66}Rh_{0.34}$	$Ag_{0.69}Rh_{0.31}$

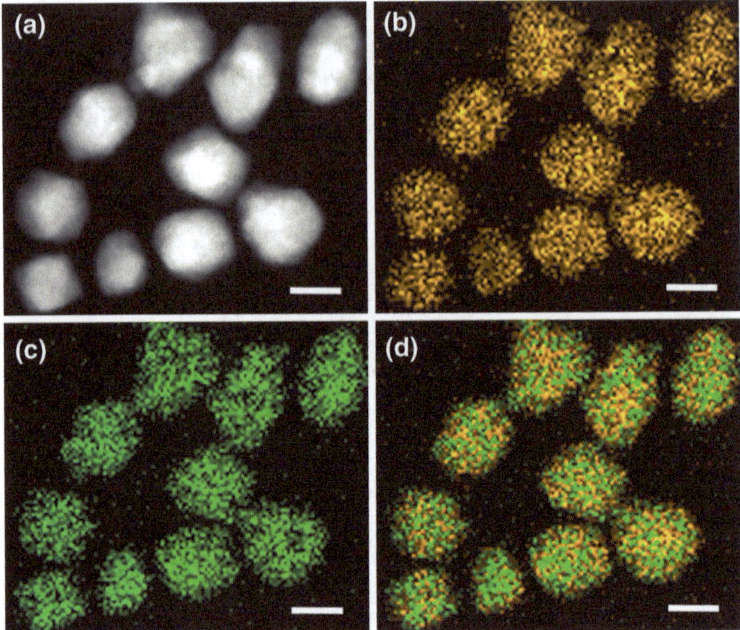

Fig. 2.3 The **a** HAADF-STEM image, **b** Ag-L STEM-EDX map and **c** Rh-L STEM-EDX map obtained from a group of prepared $Ag_{0.5}Rh_{0.5}$ nanoparticles. **d** The reconstructed overlay image of the maps shown in **b** and **c** (*green* Rh; *orange* Ag). The scale bars correspond to 10 nm (colour figure online)

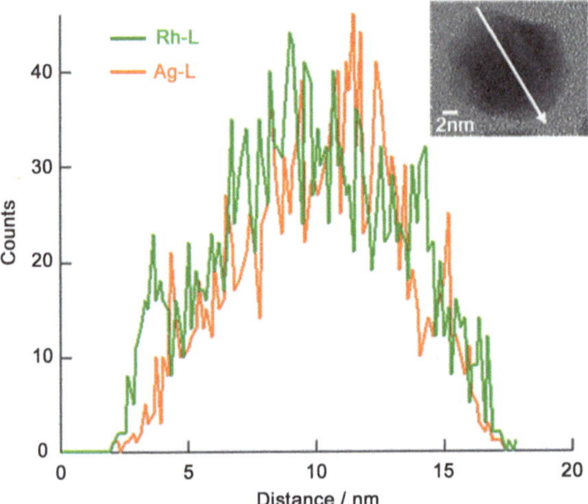

Fig. 2.4 Compositional line profiles of Ag and Rh from a $Ag_{0.5}Rh_{0.5}$ alloy nanoparticle recorded along the *arrow* shown in the STEM image (inset). Ag–L and Rh–L refer to the L electron shells of the Ag and Rh atoms, respectively. The profiles were obtained by plotting the integrated intensities at the Ag L—shell and Rh L—shell ionization edges

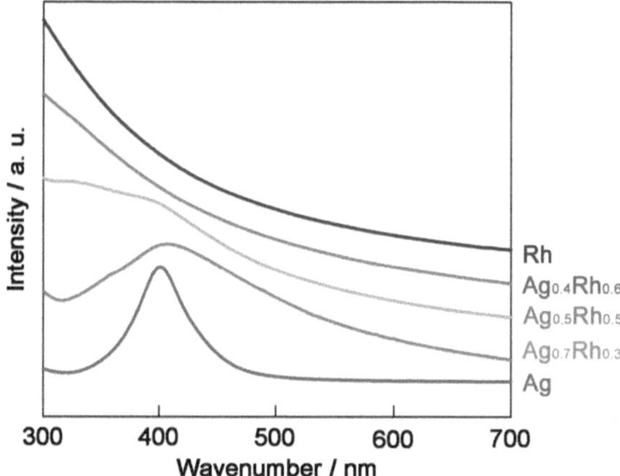

Fig. 2.5 UV-Vis absorption spectra of AgRh solid solution nanoparticles at various molar ratios

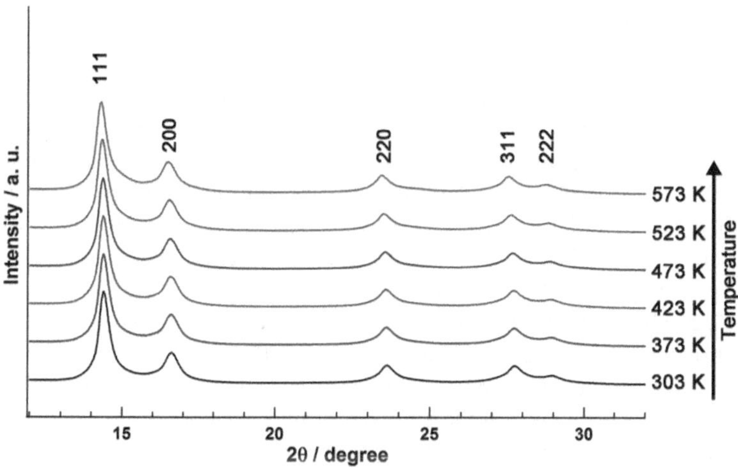

Fig. 2.6 The in situ XRD patterns of $Ag_{0.5}Rh_{0.5}$ nanoparticles measured under vacuum condition at temperature range between 303 and 573 K. The wavelength is 0.578375(1) Å

a Jasco V-570 spectrophotometer. The AgRh nanoparticles (Ag:Rh = 70:30 and 50:50) showed a broad band, in contrast to a sharp band for Ag nanoparticles and a continuous curve for AgRh nanoparticles (Ag:Rh = 40:60) and Rh nanoparticles. The surface plasmon peak observed in the solid solution alloys continuously decrease with increasing Rh content. This is caused by a continuous change of the Ag electronic state in the solid solution AgRh, indicating the formation of AgRh solid solution nanoparticles. To further investigate the thermal stability of $Ag_{0.5}Rh_{0.5}$ nanoparticles, in situ powder XRD measurements were performed at the beamline BL02B2, at SPring-8 (Fig. 2.6). It was found that the prepared $Ag_{0.5}Rh_{0.5}$ nanoparticles show the original structure up to 573 K.

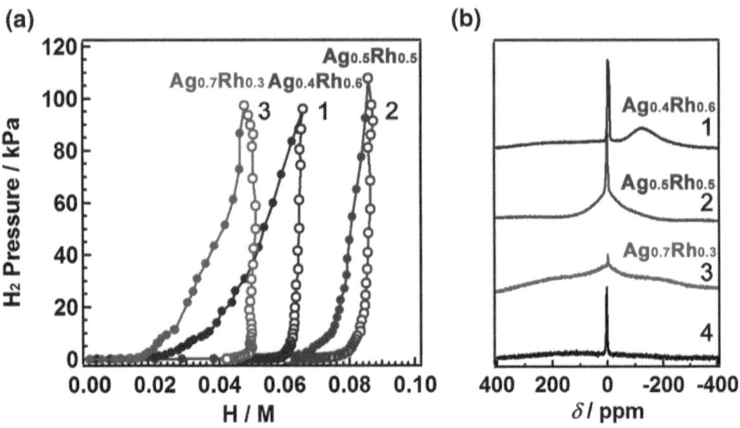

Fig. 2.7 **a** PC isotherms of *(1)* $Ag_{0.4}Rh_{0.6}$, *(2)* $Ag_{0.5}Rh_{0.5}$, and *(3)* $Ag_{0.7}Rh_{0.3}$ nanoparticles (*black circle* absorption at 303 K; *white circle* desorption at 303 K). **b** Solid-state ^2H NMR spectra for *(1)* $Ag_{0.4}Rh_{0.6}$, and *(2)* $Ag_{0.5}Rh_{0.5}$, *(3)* $Ag_{0.7}Rh_{0.3}$ nanoparticles, and *(4)* $^2H_2$ gas. All the samples were measured under 86.7 kPa of $^2H_2$ gas at 303 K

To investigate the hydrogen storage properties of the AgRh nanoparticles, hydrogen pressure-composition isotherms were measured. Studies on hydrogen storage properties give important information related to the electronic state of metals [24–28]. As shown in Fig. 2.7a, the total amount of hydrogen absorption of $Ag_{0.5}Rh_{0.5}$ nanoparticles was 0.09 H/M (M = $Ag_{0.5}Rh_{0.5}$) at ca. 100 kPa, whereas the absorptions of $Ag_{0.7}Rh_{0.3}$ and $Ag_{0.4}Rh_{0.6}$ nanoparticles were 0.05 H/M (M = $Ag_{0.7}Rh_{0.3}$) and 0.06 H/M (M = $Ag_{0.4}Rh_{0.6}$), respectively. The total amount of hydrogen absorption reached a maximum at the ratio of Ag:Rh = 50:50, where the electronic structure is expected to be similar to that of Pd. As seen in Fig. 2.7a, the amount of hydrogen absorption depends on the metal composition of the alloy. Accordingly, this difference in the amounts of hydrogen absorption implies a difference in the electronic structures of Ag_XRh_{1-X} nanoparticles.

Solid-state ^2H NMR measurements were performed to investigate the state of ^2H in the AgRh nanoparticles (Fig. 2.7b). In the spectrum of $Ag_{0.5}Rh_{0.5}$ nanoparticles, a broad absorption line with a full width at half maximum of ca. 146 ppm at ca. −2.3 ppm and a sharp line around 0 ppm were observed. In the spectrum of $^2H_2$ gas (Fig. 2.7b–4), only a sharp line at 3.4 ppm was obtained. On comparison of these spectra, it is reasonable to attribute the sharp line in the spectrum of the $Ag_{0.5}Rh_{0.5}$ particle to free deuterium gas ($^2H_2$) and the broad component to absorbed deuterium atoms (^2H) in the particles. The broad absorption lines of deuterium absorbed inside the lattices of $Ag_{0.7}Rh_{0.3}$ and $Ag_{0.4}Rh_{0.6}$ nanoparticles were observed at ca. 4.4 and −132.9 ppm, respectively (Fig. 2.7b–1, 3). These chemical shifts that give rise to the broad absorption lines suggest that the deuterium atoms inside the nanoparticles perceive the different potentials, and atomic-level alloying occurs in the Ag–Rh system.

The change in the electronic structure due to atomic-level alloying of Rh and Ag was investigated by XPS measurement (Fig. 2.8). The binding energies were

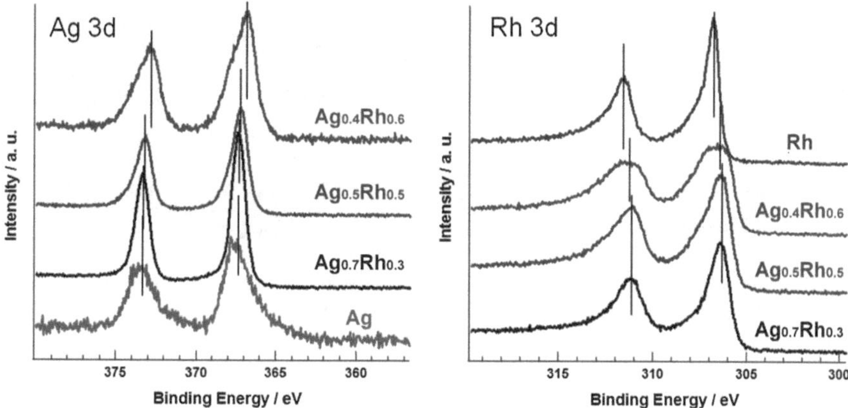

Fig. 2.8 The Ag 3d and Rh 3d core-level XPS of Ag_xRh_{1-x} nanoparticles

Table 2.2 The binding energy of the Ag 3d and Rh 3d

Sample	BE(Ag $3d_{3/2}$) (eV)	BE(Ag $3d_{5/2}$) (eV)	BE(Rh $3d_{3/2}$) (eV)	BE(Rh $3d_{5/2}$) (eV)
Rh			311.57	306.79
$Ag_{0.4}Rh_{0.6}$	372.85	366.87	311.25	306.39
$Ag_{0.5}Rh_{0.5}$	373.16	367.17	311.14	306.35
$Ag_{0.7}Rh_{0.3}$	373.32	367.30	311.14	306.36
Ag	373.38	367.39		

corrected by reference to the C(1s) line at 284.5 eV. Figure 2.8 shows Ag 3d and Rh 3d core level XPS of samples. The values of Ag 3d and Rh 3d binding energies are summarized in Table 2.2. In Ag 3d, spectral shift is different from the general shifts [29]. If silver is oxidized, the binding energy is shifted to lower energy. In Fig. 2.8, Ag 3d binding energies of alloys shifted continuously to the lower energy with increasing Rh content, which is in agreement with the oxidization of Ag. In contrast, Rh 3d binding energies of alloys shifted continuously to the lower energy with increasing Ag content, which is in agreement with the reduction of Rh. Thus the electronic structures of Rh and Ag are approached each other by the solid solution alloying.

2.4 Conclusion

In summary, the author used the chemical reduction method to obtain, for the first time, solid solutions of PVP-protected AgRh alloys that are intimately mixed at the atomic level. The atomic-level Ag–Rh alloying was confirmed by means of EDX, and XRD measurements. It is known that Ag and Rh do not mix but form segregated clustering of each element, even in the liquid phase at around 2,000 °C [9]. Consequently, solid solution AgRh alloys cannot be obtained even by means

Fig. 2.9 Graphical abstraction of hydrogen storage properties of solid solution alloys of immiscible neighboring elements with Pd

of quenching techniques, and no knowledge of the structure or the electronic structure of the alloys was available. Our results contribute to the limited knowledge in this area. Furthermore, the prepared AgRh alloys show hydrogen storage properties; the $Ag_{0.5}Rh_{0.5}$ alloy exhibits the largest amount of hydrogen absorption in the alloys. However, the H/M value of $Ag_{0.5}Rh_{0.5}$ is about half amount of Pd nanoparticles. From our experimental results, it is conclude that the $Ag_{0.5}Rh_{0.5}$ solid solution alloy has a similar electronic structure to Pd, that is to say, "modern alchemy". Following on from the discovery of the AgRh solid solution alloy, the author envisage the development of new solid solution alloys of immiscible Ag–Ni, Au–Rh, Cu–Ru, and other systems that exhibit phase-segregated structures even in high-temperature liquid phase [10, 30, 31] (Fig. 2.9).

References

1. Turchi PEA, Drchal V, Kudrnovský J (2006) Stability and ordering properties of fcc alloys based on Rh, Ir, Pd, and Pt. Phys Rev B 74:064202-1–06420212
2. Mena FP, DiTusa JF, van der Marel D, Aeppli G, Young DP, Damascelli A, Mydosh JA (2006) Suppressed reflectivity due to spin-controlled localization in a magnetic semiconductor. Phys Rev B 73:0852051–0852057
3. Eagleton TS, Mallet J, Cheng X, Wang J, Chien C, Searson PC (2005) Electrodeposition of Co_xPt_{1-x} Thin Films. J Electrochem Soc 152:C27–C31
4. Bowker M (2008) Automotive catalysis studied by surface science. Chem Soc Rev 37:2204–2211
5. Inderwildi OR, Jenkins SJ, King DA (2008) Dynamic interplay between diffusion and reaction: nitrogen recombination on Rh{211} in car exhaust catalysis. J Am Chem Soc 130:2213–2220
6. Guo J, Hsu A, Chu D, Chen R (2010) Improving oxygen reduction reaction activities on carbon-supported Ag nanoparticles in alkaline solutions. J Phys Chem C 114:4324–4330
7. Cobley CM, Campbell DJ, Xia Y (2008) Tailoring the optical and catalytic properties of gold-silver nanoboxes and nanocages by introducing palladium. Adv Mater 20:748–752
8. Holleck GL (1970) Diffusion and solubility of hydrogen in palladium and palladium-silver alloys. J Phys Chem 74:503–511
9. Zarkevich NA, Tan TL, Johnson DD (2007) First-principles prediction of phase-segregating alloy phase diagrams and a rapid design estimate of their transition temperatures. Phys Rev B 75:104203-1–10420312

10. Massalski TB, Okamoto H, Subramanian PR, Kacprzak L (1996) Binary alloy phase diagrams. ASM International
11. Gera VB, Gupta R, Jain KP (1989) Electronic structure of III-V ternary semiconductors. J Phys Condens Matter 1:4913–4930
12. Papaconstantopoulos DA, Klein BM, Economou EN, Boyer LL (1978) Band structure and superconductivity of PdD_x and PdH_x. Phys Rev B 17:141–150
13. Fazle Kibria AKM, Sakamoto Y (2000) The effect of alloying of palladium with silver and rhodium on the hydrogen solubility, miscibility gap and hysteresis. Int J Hydrogen Energy 25:53–859
14. Vuillemin JJ, Priestly MG (1965) De Haas-Van Alphen effect and fermi surface in palladium. Phys Rev Lett 14:307–309
15. Wicke E (1984) Electronic structure and properties of hydrides of 3d and 4d metals and intermetallics. J Less-Common Met 101:17–33
16. Zhou S, Jackson GS, Eichhorn B (2007) AuPt alloy nanoparticles for CO-tolerant hydrogen activation: architectural effects in Au–Pt bimetallic nanocatalysts. Adv Funct Mater 17:3099–3104
17. Hernández-Fernández P, Rojas S, Ocón P, Gómez de la Fuente JL, San Fabián J, Sanza J, Peña MA, García-García FJ, Terreros P, Fierro JLG (2007) Influence of the preparation route of bimetallic Pt–Au nanoparticle electrocatalysts for the oxygen reduction reaction. J Phys Chem C 111:2913–2923
18. Lang H, Maldonado S, Stevenson KJ, Chandler BD (2004) Synthesis and characterization of dendrimer templated supported bimetallic Pt–Au nanoparticles. J Am Chem Soc 126:12949–12956
19. Chiang I, Chen Y, Chen D (2009) Synthesis of NiAu colloidal nanocrystals with kinetically tunable properties. J Alloys Compd 468:237–245
20. Lu D, Domen K, Tanaka K (2002) Electrodeposited Au–Fe, Au–Ni, and Au–Co alloy nanoparticles from aqueous electrolytes. Langmuir 18:3226–3232
21. Chiang I, Chen D (2007) Synthesis of monodisperse FeAu nanoparticles with tunable magnetic and optical properties. Adv Funct Mater 17:1311–1316
22. Dahal N, Chikan V, Jasinski J, Leppert VJ (2008) Synthesis of water-soluble iron gold alloy nanoparticles. Chem Mater 20:6389–6395
23. Torigoe K, Nakajima Y, Esumi K (1993) Preparation and characterization of colloidal silver-platinum alloys. J Phys Chem 97:8304–8309
24. Kobayashi H, Yamauchi M, Ikeda R, Kitagawa H (2009) Atomic-level Pd–Au alloying and controllable hydrogen-absorption properties in size-controlled nanoparticles synthesized by hydrogen reduction. Chem Commun 32:4806–4808
25. Kobayashi H, Yamauchi M, Kitagawa H, Kubota Y, Kato K, Takata M (2008) Hydrogen absorption in the core/shell interface of Pd/Pt nanoparticles. J Am Chem Soc 130:1818–1819
26. Kobayashi H, Yamauchi M, Kitagawa H, Kubota Y, Kato K, Takata M (2008) On the nature of strong hydrogen atom trapping inside Pd nanoparticles. J Am Chem Soc 130:1828–1829
27. Yamauchi M, Ikeda R, Kitagawa H, Takata M (2008) Nanosize effects on hydrogen storage in palladium. J Phys Chem C 112:3294–3299
28. Yamauchi M, Kobayashi H, Kitagawa H (2009) Hydrogen storage mediated by Pd and Pt nanoparticles. Chem Phys Chem 10:2566–2576
29. Kaushik VK (1991) XPS core level spectra and auger parameters for some silver compounds. J Electron Spectrosc Relat Phenom 56:273–277
30. Liu XJ, Gao F, Wang CP, Ishida K (2008) Thermodynamic assessments of the Ag-Ni binary and Ag-Cu-Ni ternary systems. J Electron Mater 37:210–217
31. Raevskaya MV, Yanson IE, Tatarkina AL, Sokolova IG (1987) The effect of nickel on interaction in the copper ruthenium system. J Less-Common Met 132:237–241

Chapter 3
Systematic Study of the Hydrogen Storage Properties and the CO-oxidizing Abilities of Solid Solution Alloy Nanoparticles in an Immiscible Pd–Ru System

3.1 Introduction

Ru, Rh and Pd are neighboring noble metals in the 4d transition metal series. Every combination of binary alloy from these elements (Ru–Rh, Rh–Pd and Ru–Pd) does not form a solid-solution alloy throughout the whole composition range at room temperature [1–3]. Here, we have succeeded for the first time in mixing Ru and Pd, which are the elements neighboring Rh, at the atomic level over the whole composition range using chemical reduction. Solid solution alloying has the advantage of allowing researchers to discover new properties by changing compositions and/or combinations of the constituent elements. For example, although Rh and Ag have no hydrogen storage ability, $Ag_{0.5}Rh_{0.5}$ solid solution nanoparticles exhibit a hydrogen storage property like Pd, which is located between Rh and Ag in the periodic table [4]. In this chapter, the author focuses on the hydrogen storage and CO-oxidizing abilities as the first property investigation for PdRu solid solution alloy nanoparticles. As a result, the hydrogen storage property of the alloy can be tuned by controlling the composition of Pd and Ru. More noteworthy is that the CO-oxidizing ability of the alloy is extremely enhanced compared with that of Ru, Rh and Pd monometallic nanoparticles.

Palladium is well known as a hydrogen storage metal absorbing hydrogen at ambient temperature and pressure [5–9]. Many bulk metals have been investigated as hydrogen storage materials, and it was found that their hydrogen storage properties are strongly correlated with their electronic structure [10, 11]. To control the hydrogen storage properties, the influence of alloying with other metals, especially adjacent 4d metals in the periodic table, on the absorption properties has been intensively studied [12–15]. However, there are only a few reports on the hydrogen storage properties in the Pd–Ru system, despite Ru being the second nearest 4d neighbor to Pd in the periodic table [16–18].

K. Kusada, *Creation of New Metal Nanoparticles and Their Hydrogen-Storage and Catalytic Properties*, Springer Theses, DOI 10.1007/978-4-431-55087-7_3

On the other hand, Ru and Rh are well known as good catalysts; e.g., they are used as CO oxidation or NO_x reduction catalysts [19–23]. Recently, CO oxidation catalysts have been extensively developed because of their importance to CO removal from car exhaust or fuel-cell systems [24]. Pd, also well known as a valuable catalyst, is used as a catalyst for fuel-cell electrodes and exhaust-gas purification [25, 26]. However, these Pd catalysts crucially suffer from CO poisoning, which reduces catalytic activity [27], and therefore CO removal from the system is indispensable for long-term durability. Moreover, steam-reforming of shale gas is a popular source of synthesis gas for the generation of various chemical products and is strongly expected to be an important energy source in the near future. Recently, Ru has also attracted much attention as an effective catalyst for the steam-reforming reaction of methane which is the main component of shale gas [28, 29]. Furthermore, the catalytic properties of these platinum group elements are often enhanced by alloying with other elements [30–32]. Therefore, the Pd–Ru system is a promising candidate for innovative catalysts.

In addition, the 1:1 PdRu alloy is expected to possess a similar electronic structure to Rh, because Rh is located between Ru and Pd in the periodic table. Although Rh has extremely high catalytic activity, it is one of the rarest and most expensive elements. Compared with Rh, Ru and Pd are much cheaper elements, and therefore PdRu alloy is a useful material from the viewpoint of the "Element Strategy" [33]. However, there have been very few reports of PdRu solid solution alloy so far. The reason originates from the fact that Pd–Ru solid solution alloy cannot be easily obtained. Pd and Ru are immiscible at the atomic level throughout the whole composition range in the bulk state, even at high temperatures up to the melting point of Pd [3]. To obtain the PdRu solid solution alloys throughout the whole composition range, we focused on metal nanoparticles.

Recently, the nanosize effect has attracted much attention not only for developing potential applications for electronic, magnetic, optical, and catalytic materials but also for providing an efficient technique for stabilizing a nonequilibrium phase under ambient conditions [34–39]. In this chapter, the author reports the first example of Pd_xRu_{1-x} solid solution alloys over the whole composition range obtained through the chemical reduction. The hydrogen storage property of the alloy was observed, and the property was gradually changed with the composition. More noteworthy is that the CO-oxidizing ability of the alloy extremely enhanced compared to that of Ru, Rh and Pd monometallic nanoparticles.

3.2 Experiment

3.2.1 Syntheses

Syntheses of Pd_xRu_{1-x} Nanoparticles

In a typical synthesis of PdRu nanoparticles, poly(*N*-vinyl-2-pyrrolidone) (PVP, 444 mg, MW ≈ 40,000, Wako) was dissolved in triethylene glycol (TEG, 100 ml, Wako), and the solution was heated to 200 °C in air with magnetic stirring.

Table 3.1 Reaction condition for the syntheses of Pd_xRu_{1-x} nanoparticles

Sample	$K_2[PdCl_4]$ (mg)	$RuCl_3{\cdot}nH_2O$ (mg)	TEG (ml)	PVP (mmol)
$Pd_{0.1}Ru_{0.9}$	32.6	235.6	100	10.0
$Pd_{0.3}Ru_{0.7}$	98.0	180.3	100	1.0
$Pd_{0.5}Ru_{0.5}$	163.4	131.1	100	1.0
$Pd_{0.7}Ru_{0.3}$	228.7	62.4	100	1.0
$Pd_{0.9}Ru_{0.1}$	293.8	25.9	100	1.0

Meanwhile, $K_2[PdCl_4]$ (163.4 mg, Aldrich) and $RuCl_3{\cdot}nH_2O$ (131.1 mg, Wako) were dissolved in deionized water (40 ml). The aqueous mixture solution of $K_2[PdCl_4]$ and $RuCl_3{\cdot}nH_2O$ was then added to the triethylene glycol. The solution was maintained at 200 °C while adding the solution. After the reaction was complete, the prepared nanoparticles were separated by centrifugation.

Other Pd_xRu_{1-x} (x = 0.1, 0.3, 0.7 and 0.9) nanoparticles were prepared by controlling the molar ratio of Pd^{2+} and Ru^{3+} ions. The details of the synthesis conditions for Pd_xRu_{1-x} nanoparticles are summarized in Table 3.1.

Synthesis of Pd Nanoparticles

In a synthesis of Pd nanoparticles, PVP (555 mg, MW ≈ 40,000, Wako) was dissolved in TEG (100 ml, Wako), and the solution was heated to 200 °C in air with magnetic stirring. Meanwhile, $K_2[PdCl_4]$ (326.3 mg, Aldrich) was dissolved in deionized water (40 ml). The aqueous solution of $K_2[PdCl_4]$ was then added to the triethylene glycol. Then the solution was maintained at 200 °C while adding the solution. After the reaction was complete, the prepared nanoparticles were separated by centrifugation.

Synthesis of Ru Nanoparticles

In a synthesis of Ru nanoparticles, $RuCl_3{\cdot}nH_2O$ (783.2 mg) and PVP (111 mg, MW ≈ 40,000, Wako) were dissolved in TEG (50 ml) at room temperature. Then the solution was heated to 200 °C and maintained at this temperature for 6 h. After the reaction was complete, the prepared nanoparticles were separated by centrifugation.

3.2.2 Catalysts Preparation

To investigate the CO-oxidizing catalytic activity, Ru, Rh, Pd and Pd_xRu_{1-x} alloy nanoparticles and a physical mixture (Ru and Pd nanoparticles) supported on γ-Al_2O_3 catalysts were prepared by wet impregnation. Each nanoparticle (equivalent to 1 wt% of γ-Al_2O_3) was ultrasonically dispersed in purified water. γ-Al_2O_3

support that had been precalcined at 1,073 K for 5 h was put in each aqueous solution of nanoparticles and then the suspended solutions were stirred for 12 h. After stirring, the suspended solutions were heated to 60 °C and dried under vacuum. The resulting powders were kept at 120 °C for 8 h to remove water completely.

3.2.3 Characterizations

Transmission Electron Microscopy Analysis

Transmission electron microscopy (TEM) images were recorded on a Hitachi HT7700 TEM instrument operated at 100 kV accelerate voltage. The samples dispersed with ethanol were drop-cast onto a carbon-coated copper grid and allowed to dry under ambient conditions.

High Resolution TEM and Selected Area Diffraction Pattern Analyses

High resolution TEM (HRTEM), selected area diffraction pattern (SADP) and dark field (DF) TEM images were captured using a JEOL JEM-3200FSK TEM instrument operated at 300 kV. The samples dispersed with ethanol were drop-cast onto a carbon-coated copper grid and allowed to dry under ambient conditions.

Synchrotron X-ray Diffraction Measurement

Synchrotron X-ray diffraction (XRD) patterns were measured at the beam line BL02B2, SPring-8. The diffraction patterns of the samples sealed in glass capillaries under vacuum were measured at 303 K. The wavelength is $\lambda = 0.55277$ Å.

Scanning Transition Electron Microscopy Analysis

HRSTEM, high-angle annular dark-field (HAADF) scanning transition electron microscopy (STEM) and energy-dispersive X-ray (EDX) analyses were recorded on a HITACHI HD-2700 STEM operated at 200 kV accelerate voltage at the Naka Application Center, Hitachi High-Technologies Corporation and JEM-ARM200F operated at 200 kV accelerate voltage.

In Situ XRD Measurement

The thermal stability of $Pd_{0.5}Ru_{0.5}$ nanoparticles were investigated by in situ powder XRD analysis measured at the BL02B2 beamline, SPring-8. The XRD patterns of the samples sealed in a glass capillary were measured in situ under vacuum condition at temperature range between 303 and 723 K. The wavelength is 0.578375(1) Å.

Pressure-Composition Isotherms Measurement

Pressure-Composition (PC) isotherms were measured at hydrogen pressure range between 0 and ca. 100 kPa at 303 K by a volumetric technique using a pressure-composition-temperature apparatus (Suzuki Shokan Co., Ltd).

Solid-State ^2H NMR Spectra Measurement

Solid-state ^2H NMR spectra were recorded at 303 K in a fixed magnetic field of 94 kOe and a frequency sweep of 60.94–61.94 MHz, using a BRUKER NMR spectrometer. Each sample was evacuated in a glass capillary for several hours at 373 K, and then each sample was sealed into a glass capillary with 101.3 kPa of ^2H gas at 303 K.

X-ray Photoelectron Spectroscopy Analysis

X-ray photoelectron spectroscopy (XPS_ spectra for samples on an carbon sheet were recorded on a Shimadzu ESCA-3400 X-ray photoelectron spectrometer.

Catalytic Test

The obtained catalyst powders were pressed into pellets at 1.2 MPa for 5 min. The pellets were crushed and sieved to obtain grains with diameters between 180 and 250 μm. Each supported nanoparticle catalyst (150 mg) was loaded into a tubular quartz reactor (i.d. 7 mm) with quartz wool. $CO/O_2/He$ mixed gas ($He/CO/O_2$: 49/0.5/0.5 ml min^{-1}) was passed over the catalysts at ambient temperature, and the catalysts were then heated to 100 °C. After 15 min, effluent gas was collected, and the reaction products were analyzed by gas chromatography with a thermal conductivity detector (GC-8A, Shimadzu, Japan). Catalysts were heated in increments of 10 °C to a temperature at which CO was consumed completely, and the products were analyzed at each temperature. After the reaction, the reactor was purged with He at the reaction temperature, and the catalysts were then cooled to room temperature.

3.3 Results and Discussion

To investigate the composition of Pd and Ru atoms in the nanoparticles, elemental analyses were carried out using EDX techniques. The average stoichiometries determined from the EDX data were summarized in Table 3.2 and Fig. 3.1. TEM images of the synthesized Pd_xRu_{1-x} nanoparticles were recorded on a Hitachi HT7700 TEM instrument (Fig. 3.2). The mean diameters of the nanoparticles

Table 3.2 Results of the metal composition of PdRu bimetallic nanoparticles

Sample	Pd^+:Ru^{3+} (nominal composition)	Pd:Ru (determined from EDX)
$Pd_{0.1}Ru_{0.9}$	1:9	12.8:87.2
$Pd_{0.3}Ru_{0.7}$	3:7	29.7:70.3
$Pd_{0.5}Ru_{0.5}$	5:5	47.9:52.1
$Pd_{0.7}Ru_{0.3}$	7:3	73.6:26.4
$Pd_{0.9}Ru_{0.1}$	9:1	90.9:9.1

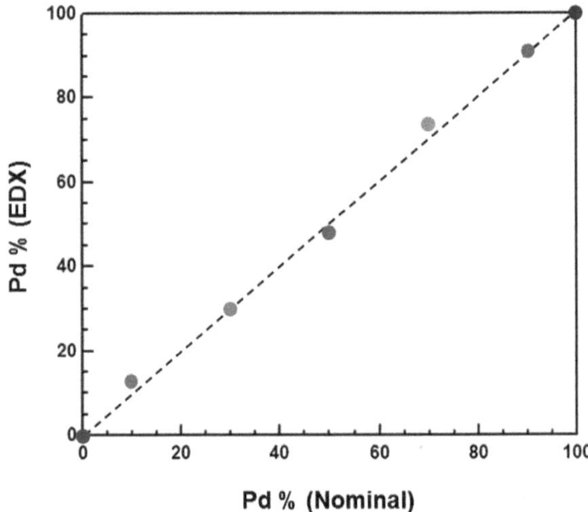

Fig. 3.1 The plot of estimated composition (determined from EDX) versus nominal composition

were determined from the TEM images to be (a) 6.4 ± 1.7, (b) 9.4 ± 1.7, (c) 12.5 ± 2.2, (d) 10.0 ± 1.2, (e) 8.2 ± 1.6, (f) 8.6 ± 1.4 and (g) 9.8 ± 2.6, respectively. Numbers that follow the ± sign represent estimated standard deviations. These mean diameters were estimated by averaging over at least 300 particles. Nonspherical nanoparticles were observed in Ru-rich nanoparticles; this is considered to arise from the nature of hcp-structured Ru with ab anisotropic growth direction as shown in Fig. 3.2a.

First of all, to clarify the structure of the prepared PdRu nanoparticles, $Pd_{0.5}Ru_{0.5}$ nanoparticles were investigated in detail. Figure 3.3 shows elemental mapping data for the prepared $Pd_{0.5}Ru_{0.5}$ nanoparticles. Figure 3.3a is a HAADF STEM image. Figure 3.3b and c are the corresponding Pd–L and Ru–L STEM–EDX maps, respectively. Figure 3.3d is an overlay map of the Pd and Ru chemical distributions. The map provides visual evidence of the formation of $Pd_{0.5}Ru_{0.5}$ solid solutions. The author further characterized the $Pd_{0.5}Ru_{0.5}$ nanoparticles by EDX line scanning analysis (Fig. 3.3e, f). The line-scan position of the nanoparticle is denoted by the white arrow in Fig. 3.3e. The compositional line profiles of Pd and Ru on a $Pd_{0.5}Ru_{0.5}$ nanoparticle show that Ru and Pd atoms are homogeneously distributed over the whole particle.

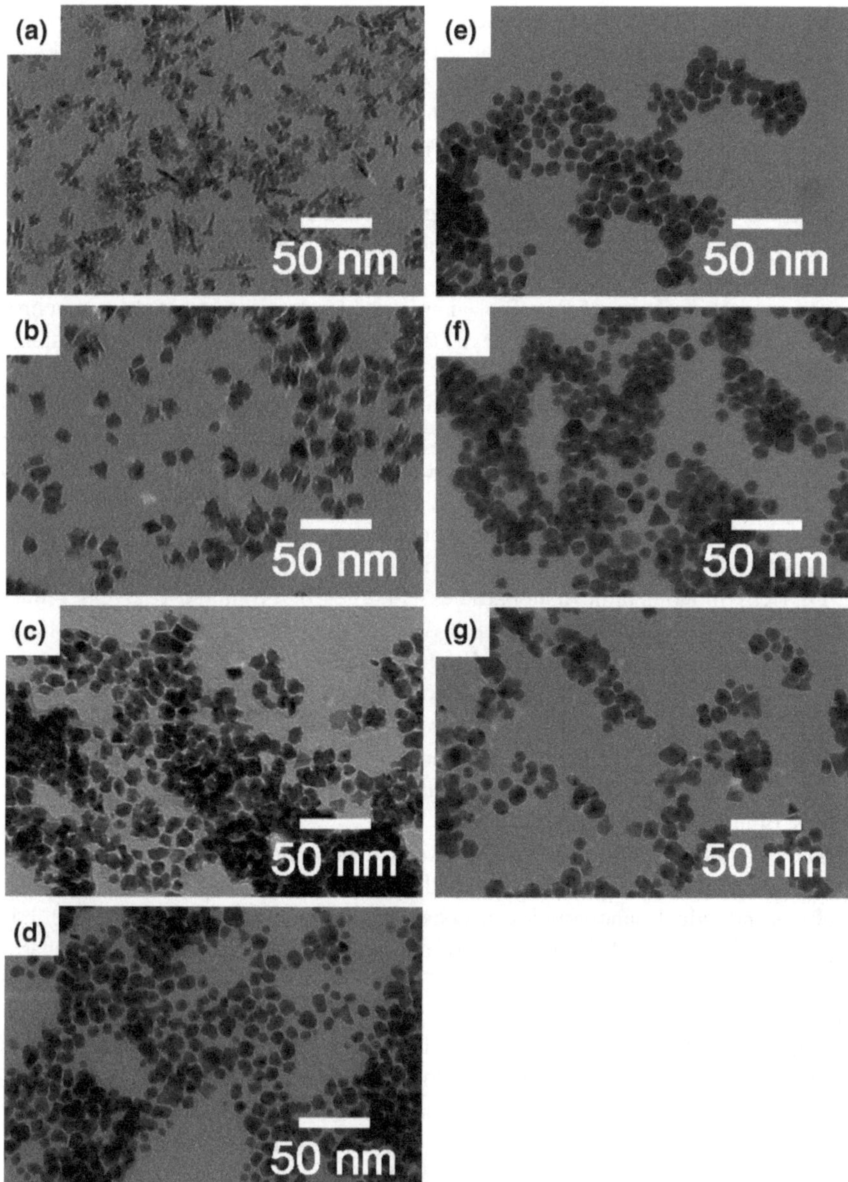

Fig. 3.2 The TEM images of **a** Ru, **b** $Pd_{0.1}Ru_{0.9}$, **c** $Pd_{0.3}Ru_{0.7}$, **d** $Pd_{0.5}Ru_{0.5}$, **e** $Pd_{0.7}Ru_{0.3}$, **f** $Pd_{0.9}Ru_{0.1}$, and **g** Pd nanoparticles, respectively

These results indicate the formation of atomic-level PdRu alloy. Pd and Ru have different redox potentials i.e. Pd ion is easily reduced to Pd metal compared with Ru ion. If both Pd and Ru precursors are added to TEG solution before heating, Pd ions are firstly reduced around 90 °C and Ru ions are subsequently reduced around 160 °C by TEG, resulting in the formation of the segregated structure of Pd and Ru. The point of synthesis for solid-solution structure is to slowly add these metal precursors into thoroughly heated TEG solution to simultaneously reduce these metal ions to metals. In this report, the metal precursor solution slowly added to TEG heated at 200 °C. The added metal solution turned black in a moment, indicating that both metal ions are quickly reduced and the difference of reduction speed is considered to be negligible. As a consequence of the simultaneous reduction for Pd and Ru ions, we successfully synthesized the solid-solution structure nanoparticles.

The crystal structure of $Pd_{0.5}Ru_{0.5}$ nanoparticles was investigated by synchrotron XRD analysis at the beamline BL02B2, SPring-8. The diffraction patterns of the samples sealed in glass capillaries under vacuum were measured at 303 K. Figure 3.4a and b show the XRD patterns of Ru, Pd and PdRu nanoparticles. As shown in Fig. 3.4a, Ru and Pd nanoparticles showed a powder XRD pattern originating from a single hexagonal close-packed (hcp) and face-centered cubic (fcc) lattice, respectively, as well as those bulks. By contrast, $Pd_{0.5}Ru_{0.5}$ nanoparticles have an XRD pattern comprising two components diffracted from fcc and hcp lattices, where the positions of the diffraction peaks differ from those of Ru (hcp) or Pd (fcc) monometallic nanoparticles (Fig. 3.4b). From the Rietveld refinement of the diffraction pattern for $Pd_{0.5}Ru_{0.5}$ nanoparticles, the lattice constants for the two components were refined to be $a = 3.852(1)$ Å for the fcc lattice and $a = 2.7227(8)$ Å and $c = 4.381(2)$ Å for the hcp lattice (Fig. 3.4c). The lattice constant of the fcc component is smaller than that of Pd ($a = 3.8925$ Å), and the lattice constants of the hcp components are larger than those of Ru ($a = 2.7056$ Å, $c = 4.2793$ Å). These results strongly support the formation, in both fcc and hcp phases, of atomic-level $Pd_{0.5}Ru_{0.5}$ alloys.

The structure of solid solution $Pd_{0.5}Ru_{0.5}$ nanoparticles was further characterized by TEM; i.e., the author determined whether fcc and hcp phases exist separately as individual nanoparticles or coexist in a single nanoparticle. Figure 3.5a shows a bright field (BF)TEM image of a $Pd_{0.5}Ru_{0.5}$ nanoparticle. The nanoparticle is polycrystalline with small grains. Figure 3.5b is the SADP of the $Pd_{0.5}Ru_{0.5}$ nanoparticle in Fig. 3.5a. The d-spacings from diffraction spots A and B were calculated to be 2.24 and 1.69 Å, respectively. Taking into account d-values obtained from the XRD pattern, the diffraction spots of A and B correspond to fcc (111) (d-spacing of 2.24 Å) and hcp (012) (d-spacing of 1.61 Å) lattices, respectively.

Figures 3.5c and d are dark-field (DF)TEM images obtained for diffraction spots A and B, respectively. Figure 3.5c and d show fcc (111) and hcp (012) grains, respectively, and it is seen that each grain is located in a different position. These results clearly demonstrate that the fcc and hcp phases coexist in a single $Pd_{0.5}Ru_{0.5}$ solid solution nanoparticle.

Nanoparticles of the other compositions were also investigated by similar methods, including elemental mapping and XRD measurements.

Figure 3.6 shows elemental mapping data for the prepared Pd_xRu_{1-x} nanoparticles. Figure 3.6a, c, e and g are HAADF-STEM images of Pd_xRu_{1-x} nanoparticles. Figure 3.6b, d, f and h are the overlay maps of the corresponding Pd–L (blue)

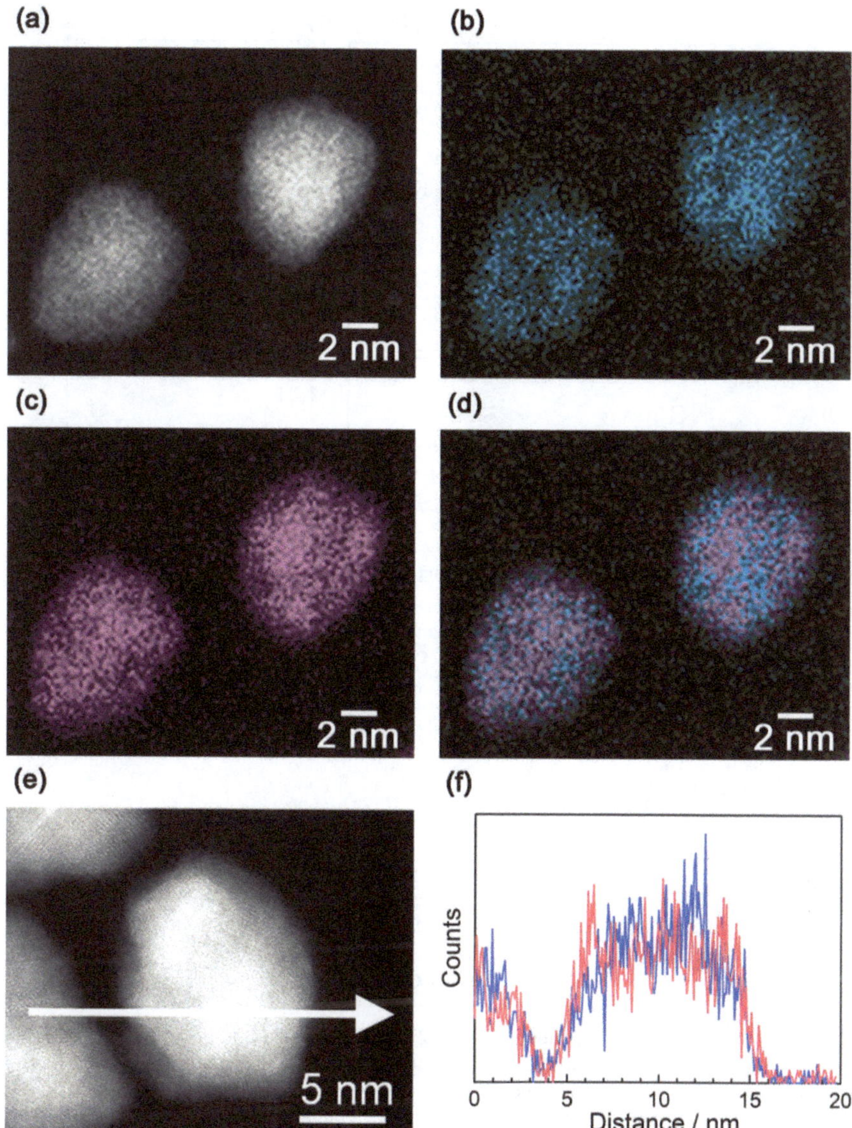

Fig. 3.3 **a** HAADF–STEM image, **b** Pd–L STEM–EDX map and **c** Ru–L STEM–EDX map obtained for a group of prepared $Pd_{0.5}Ru_{0.5}$ nanoparticles. **d** Reconstructed overlay image of the maps shown in panels b and c (*blue* Pd; *red* Ru). **f** Compositional line profiles of Pd (*blue*) and Ru (*red*) for the $Pd_{0.5}Ru_{0.5}$ nanoparticle recorded along the *arrow* shown in the STEM image **e**

and Ru–L (red) STEM–EDX maps of Pd_xRu_{1-x} nanoparticles. These maps provide visual evidence of the formation of Pd_xRu_{1-x} solid solutions.

The crystal structures of Pd_xRu_{1-x} bimetallic nanoparticles were also investigated by synchrotron XRD at the beam line BL02B2, SPring-8 ($\lambda = 0.57803$ Å),

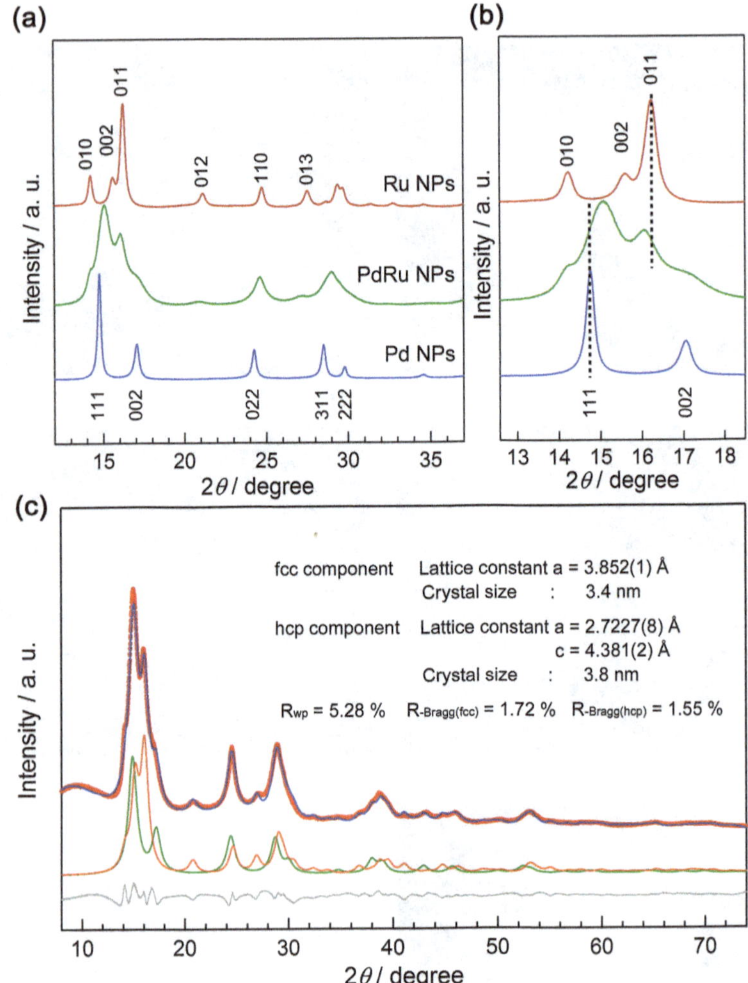

Fig. 3.4 **a** Synchrotron XRD patterns ($2\theta = 12°–37°$) of Ru, Pd and $Pd_{0.5}Ru_{0.5}$ nanoparticles at 303 K; **b** close-up of the $2\theta = 12.5°–19°$ region. **c** The diffraction pattern of $Pd_{0.5}Ru_{0.5}$ nanoparticles (*red circle*) at 303 K and calculated pattern (*blue line*). The *bottom lines* show the difference profile (*gray*) and the fitting *curves* of the fcc component (*green*) and hcp component (*orange*). The radiation wavelength is 0.57803(2) Å

as with $Pd_{0.5}Ru_{0.5}$ nanoparticles. Figure 3.7a shows the XRD patterns of the Pd and Ru, and Pd_xRu_{1-x} nanoparticles. While the Pd nanoparticles showed the fcc diffraction pattern, the Ru nanoparticles showed the hcp diffraction pattern. The dominant diffraction patterns of Pd_xRu_{1-x} nanoparticles change from the fcc pattern to the hcp pattern with increasing Ru content in the PdRu nanoparticles. The crystal structures of Pd_xRu_{1-x} were also determined by the Rietveld refinement. The refinement results were summarized in Figs. 3.8, 3.9, 3.10, 3.11, 3.12

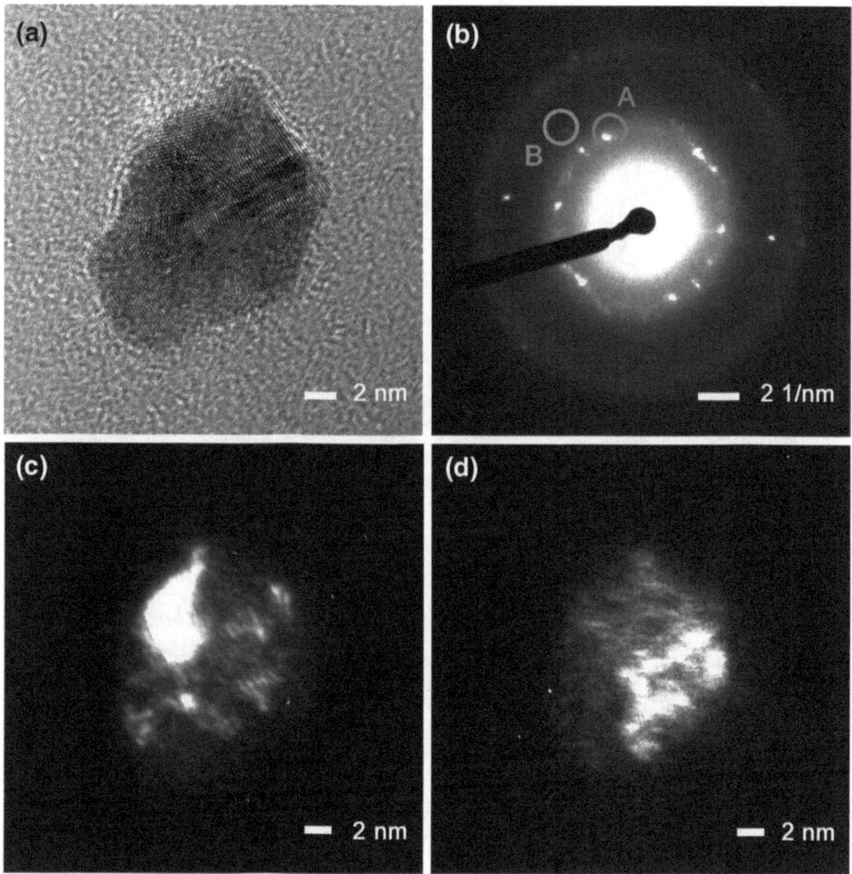

Fig. 3.5 **a** BFTEM image of a $Pd_{0.5}Ru_{0.5}$ nanoparticle, **b** SADP of (**a**), **c** DFTEM image of a $Pd_{0.5}Ru_{0.5}$ nanoparticle. Part of the fcc(111) diffraction spots was selected as the image-forming diffraction vector, as indicated schematically by A (*green circle*) in (**b**). **d** DFTEM image of a $Pd_{0.5}Ru_{0.5}$ nanoparticle. Part of the hcp(012) diffraction spots was selected as the image-forming diffraction vector, as indicated schematically by B (*orange circle*) in (**b**)

and 3.13. From the refinement results, the structures of Pd_xRu_{1-x} nanoparticles ($0.3 \leq x \leq 0.7$) consisted of both fcc and hcp structures.

To further investigate the structure of alloy nanoparticles, the change of the lattice parameter was focused on as a function of metal compositions. Figure 3.7b shows the lattice constants of Pd_xRu_{1-x} nanoparticles estimated from the Rietveld refinements of the XRD patterns. Because both of the fcc and hcp structures are close-packed structures, the square root of the lattice parameter a in the hcp component is in accordance with the lattice parameter a in the fcc component (See in Fig. 3.14). As shown in Fig. 3.7b, the lattice constants increased linearly with increasing the Pd content (x). The linear correlation between the lattice constant

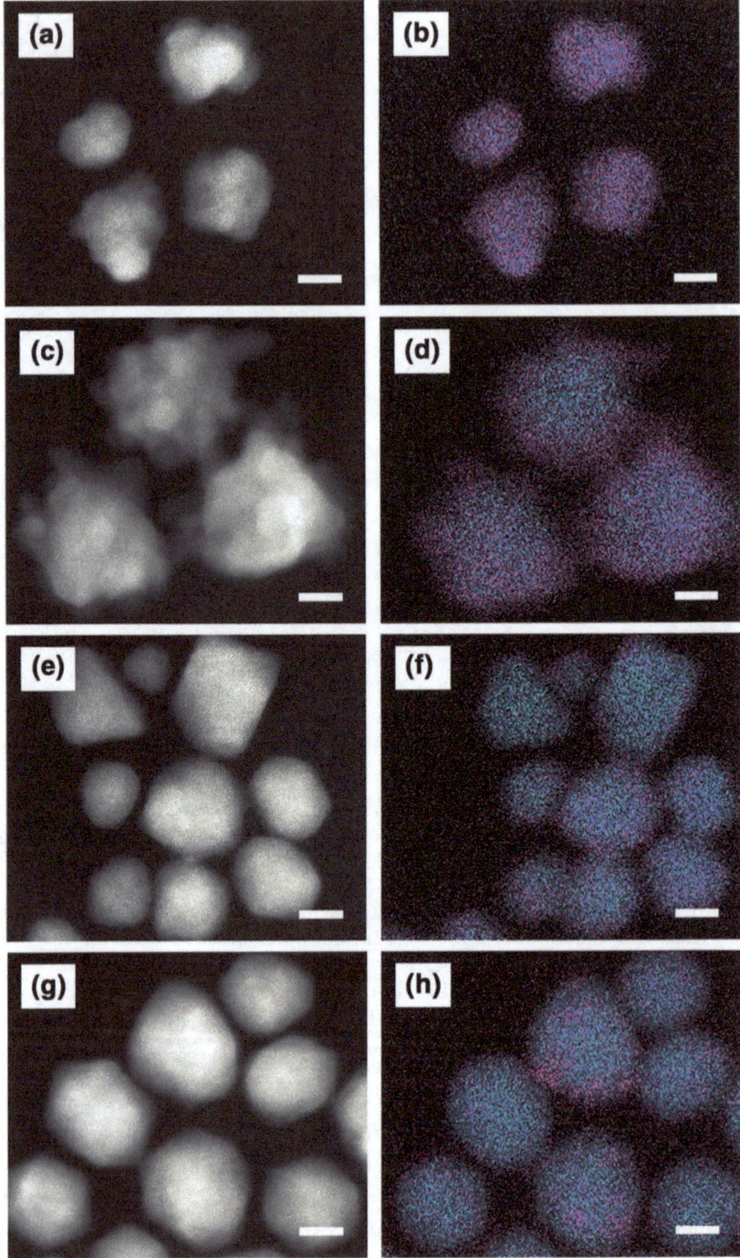

Fig. 3.6 The HAADF–STEM images of **a** $Pd_{0.1}Ru_{0.9}$, **c** $Pd_{0.3}Ru_{0.7}$, **e** $Pd_{0.7}Ru_{0.3}$, and **g** $Pd_{0.9}Ru_{0.1}$; The reconstructed overlay images of Pd–L STEM–EDX maps and Ru–L STEM–EDX maps obtained for groups of prepared Pd_xRu_{1-x} nanoparticles (*blue* Pd; *red* Ru) (**b**) $Pd_{0.1}Ru_{0.9}$, **d** $Pd_{0.3}Ru_{0.7}$, **f** $Pd_{0.7}Ru_{0.3}$, and **h** $Pd_{0.9}Ru_{0.1}$

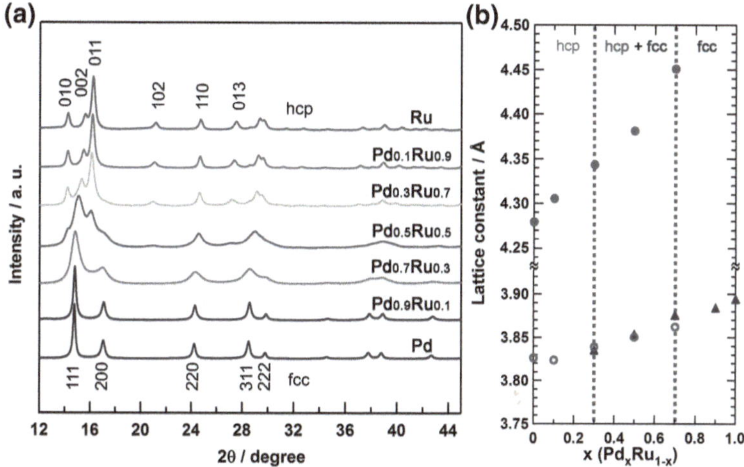

Fig. 3.7 **a** The synchrotron XRD patterns ($2\theta = 12°–45°$) of Pd_xRu_{1-x} nanoparticles at 303 K. The radiation wavelength is 0.57803(2) Å. **b** The dependence of the lattice constant on the metal composition in Pd_xRu_{1-x} nanoparticles at 303 K. *Circle* (*red*) the lattice constant a of the hcp component, *filled circle* (*red*) the lattice constant c of the hcp component, *filled up pointed triangle* (*blue*) the lattice constant of the fcc component (color figure online)

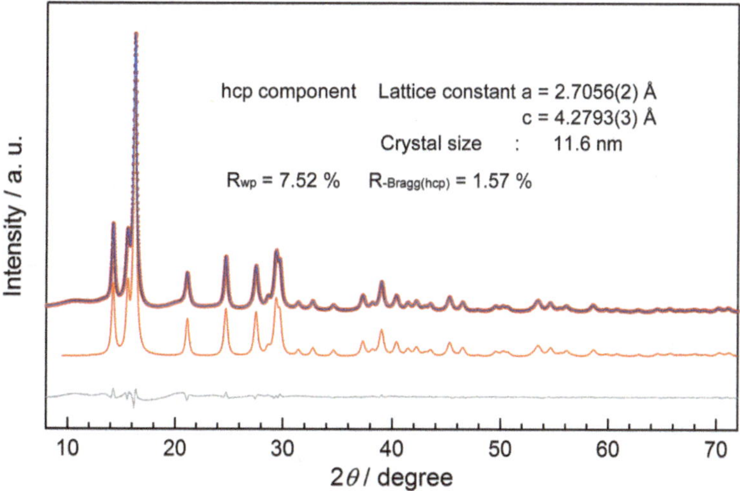

Fig. 3.8 The diffraction pattern of Ru nanoparticles (*red circle*) at 303 K and calculated pattern (*blue line*). The *bottom lines* show the difference profile (*gray*) and the fitting curves of the hcp component (*orange*)

and the alloy composition follows Vegard's law, which is an approximate empirical rule proposing a linear relation, at constant temperature, between the crystal lattice constant of an alloy and the concentrations of the constituent elements [40]. Furthermore in the Pd composition range of 30–70 %, the lattice constants a in fcc

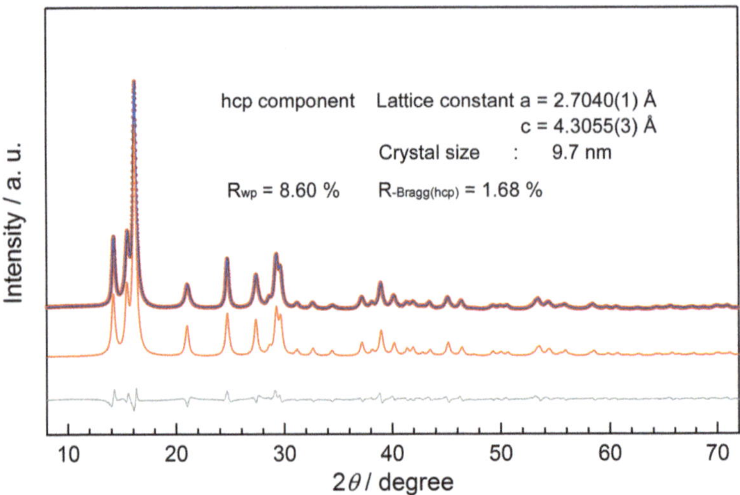

Fig. 3.9 The diffraction pattern of $Pd_{0.1}Ru_{0.9}$ nanoparticles (*red circle*) at 303 K and calculated pattern (*blue line*). The *bottom lines* show the difference profile (*gray*) and the fitting *curves* of the hcp component (*orange*)

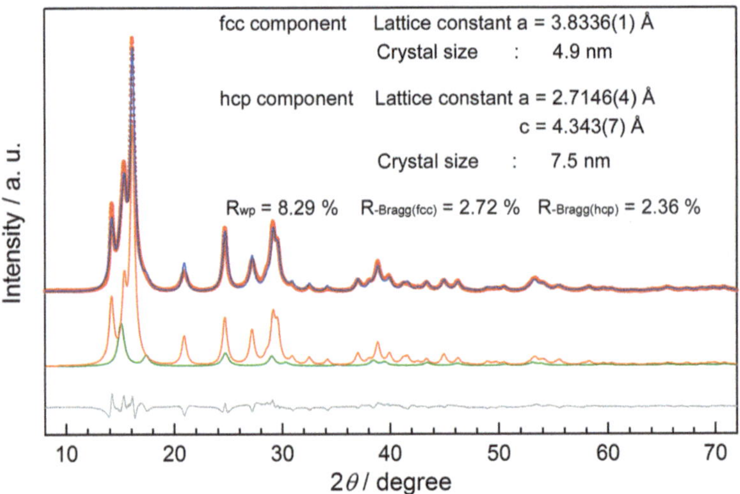

Fig. 3.10 The diffraction pattern of $Pd_{0.3}Ru_{0.7}$ nanoparticles (*red circle*) at 303 K and calculated pattern (*blue line*). The *bottom lines* show the difference profile (*gray*) and the fitting *curves* of the fcc component (*green*) and hcp component (*orange*)

and the square root of the lattice constant a in hcp components at the same metal composition have almost same value. This result means that the metal compositions in coexisting fcc and hcp phases have almost same values, and these values correspond to the average stoichiometry determined from the EDX data. These

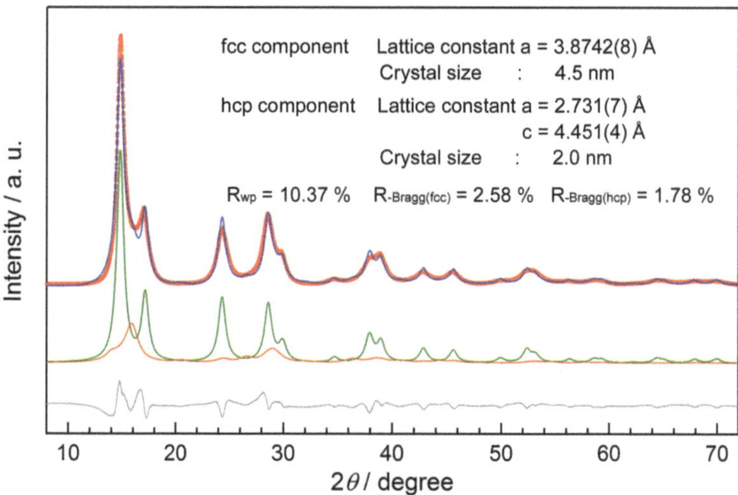

Fig. 3.11 The diffraction pattern of $Pd_{0.7}Ru_{0.3}$ nanoparticles (*red circle*) at 303 K and calculated pattern (*blue line*). The *bottom lines* show the difference profile (*gray*) and the fitting *curves* of the fcc component (*green*) and hcp component (*orange*)

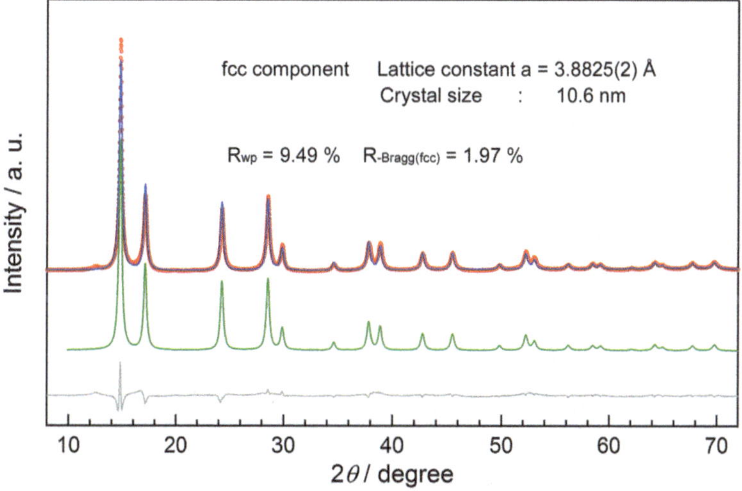

Fig. 3.12 The diffraction pattern of $Pd_{0.9}Ru_{0.1}$ nanoparticles (*red circle*) at 303 K and calculated pattern (*blue line*). The *bottom lines* show the difference profile (*gray*) and the fitting curves of the fcc component (*green*)

results strongly support the formation of the atomic-level PdRu alloy over the whole composition.

To further investigate the thermal stability of $Pd_{0.5}Ru_{0.5}$ nanoparticles, in situ powder XRD measurements were performed at the beamline BL02B2, at SPring-8

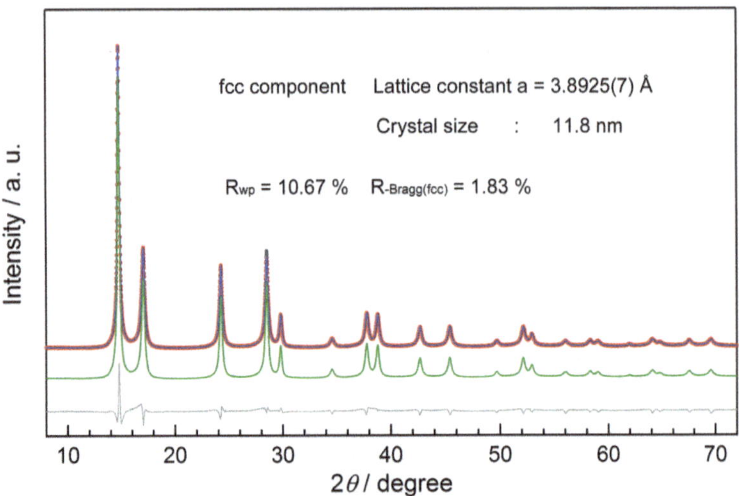

Fig. 3.13 The diffraction pattern of Pd nanoparticles (*red circles*) at 303 K and calculated pattern (*blue line*). The *bottom lines* show the difference profile (*gray*) and the fitting *curves* of the fcc component (*green*)

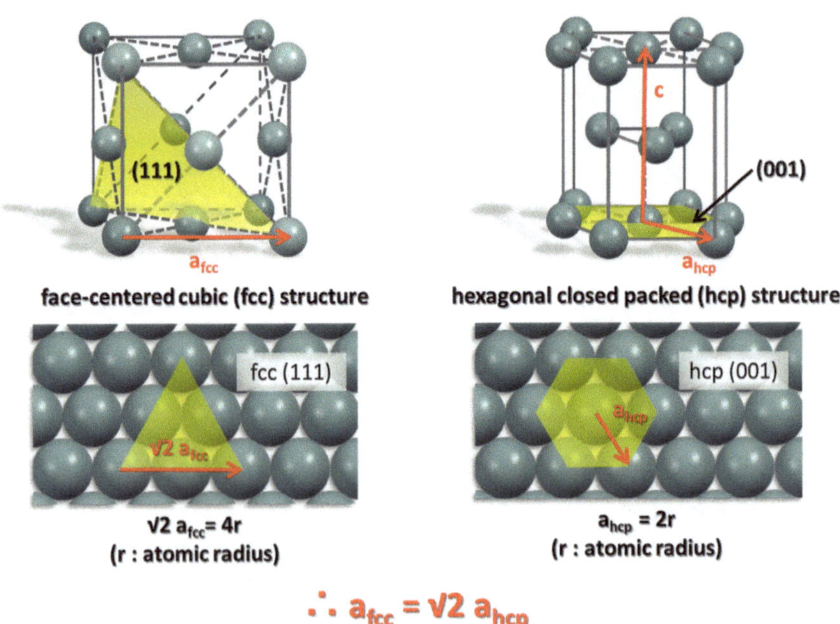

Fig. 3.14 The comparison of the fcc with hcp structures

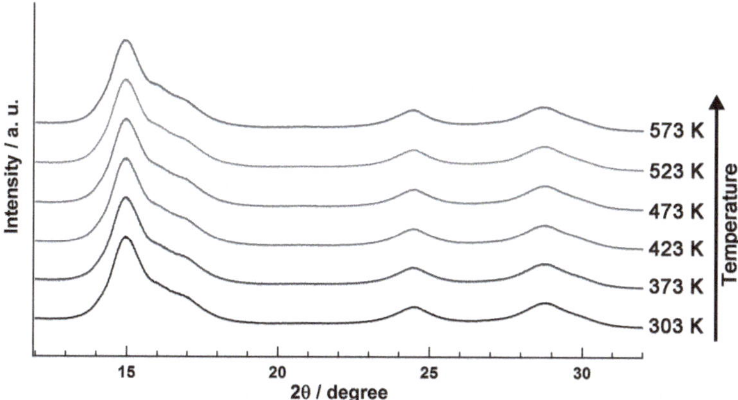

Fig. 3.15 The in situ XRD patterns of $Pd_{0.5}Ru_{0.5}$ nanoparticles measured under vacuum condition at temperature range between 303 and 573 K. The wavelength is 0.578375(1) Å

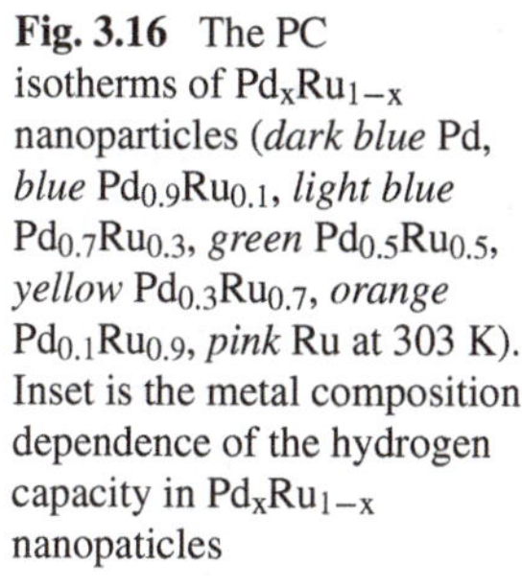

Fig. 3.16 The PC isotherms of Pd_xRu_{1-x} nanoparticles (*dark blue* Pd, *blue* $Pd_{0.9}Ru_{0.1}$, *light blue* $Pd_{0.7}Ru_{0.3}$, *green* $Pd_{0.5}Ru_{0.5}$, *yellow* $Pd_{0.3}Ru_{0.7}$, *orange* $Pd_{0.1}Ru_{0.9}$, *pink* Ru at 303 K). Inset is the metal composition dependence of the hydrogen capacity in Pd_xRu_{1-x} nanopaticles

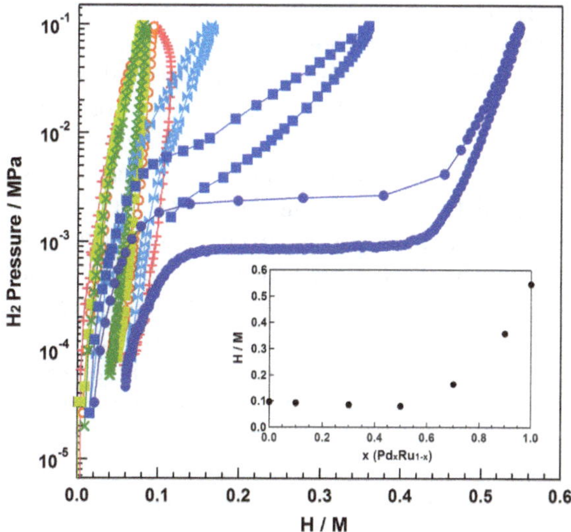

(Fig. 3.15). It was found that the prepared $Pd_{0.5}Ru_{0.5}$ nanoparticles show the original structure up to 573 K.

Studies on hydrogen-absorption properties of the metal nanoparticles give important information related to the structure and the electronic state [10, 11]. To investigate the hydrogen-absorption properties accompanied by the addition of Ru atoms to Pd nanoparticles, PC isotherms of Pd_xRu_{1-x} nanoparticles were measured at 303 K. As shown in Fig. 3.16, the hydrogen concentration of PdRu nanoparticles decreased with increasing Ru content. 30 at.% replacement of Pd with Ru results in a reduction of more than half of the total amount of hydrogen absorption

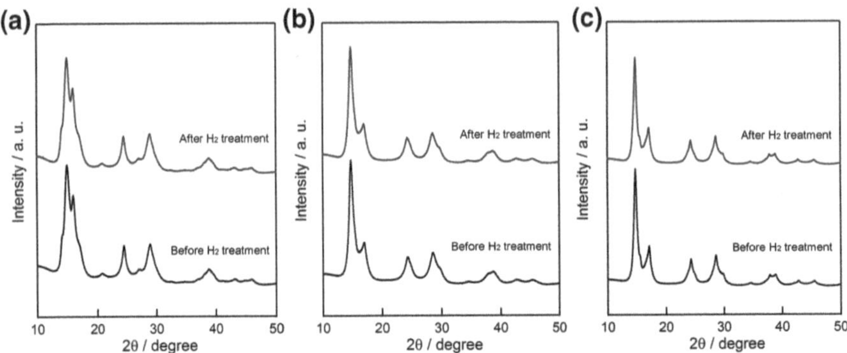

Fig. 3.17 The synchrotron in situ XRD patterns of **a** $Pd_{0.5}Ru_{0.5}$, **b** $Pd_{0.7}Ru_{0.3}$ and **c** $Pd_{0.9}Ru_{0.1}$ nanoparticles. The radiation wavelength was 0.57803(2) Å

at ca. 100 kPa (0.2 $H/Pd_{0.7}Ru_{0.3}$). With the further addition of Ru atoms, the amount of absorbed hydrogen decreased to 0.1 H/M below $x = 0.5$ (Fig. 3.16 inset). This drastic reduction in hydrogen absorption suggests a change in the thermodynamic behavior for hydrogen storage in PdRu nanoparticles with metal composition. It was confirmed that the solid-solution structures of Pd-Ru nanoparticles are maintained before and after hydrogen treatment by XRD measurements (Fig. 3.17).

Figure 3.18 shows the temperature dependences of PC isotherms in Pd_xRu_{1-x} nanoparticles. As shown in Fig. 3.18a, Pd nanoparticles exhibited an exothermic reaction with hydrogen analogously to Pd bulk [5–9], i.e., Pd nanoparticles absorb hydrogen at 303 K more than at 423 K. The reaction of hydrogen with $Pd_{0.9}Ru_{0.1}$ nanoparticles was also exothermic. The tendency of temperature dependence of PC isotherms further changes with metal composition and finally the $Pd_{0.5}Ru_{0.5}$ nanoparticles can absorb more hydrogen at 423 K than at 303 K. These results show that the enthalpy of hydrogen absorption for the Pd_xRu_{1-x} nanoparticles varied from exothermic to endothermic as the Ru content increased. This phenomenon might be caused by the change in electronic state of Pd with increasing Ru. The significance of this work is the success in continuously controlling the thermodynamic parameters for the reaction of Pd_xRu_{1-x} with hydrogen by atomic-level alloying.

To investigate the further hydrogen storage properties, the numerical values of the thermodynamic parameters of hydrogen absorption in Pd_xRu_{1-x} nanoparticles were estimated. In the hydrogen storage metals exhibiting an exothermic reaction with hydrogen, there are two phases through the reaction with hydrogen. One is the solid solution phase (α phase; M + H) of metals and hydrogen and the other is the hydride phase (β phase; M − H) with a metallic bond. The phase transition from α phase to β phase takes place with an accompanying plateau pressure in the PC isotherm. From Fig. 3.16, plateau-like behavior was observed in PC isotherms of Pd, $Pd_{0.9}Ru_{0.1}$ and $Pd_{0.7}Ru_{0.3}$ nanoparticles. This implies that these three kinds of nanoparticles were transformed from α phase to β phase and the other

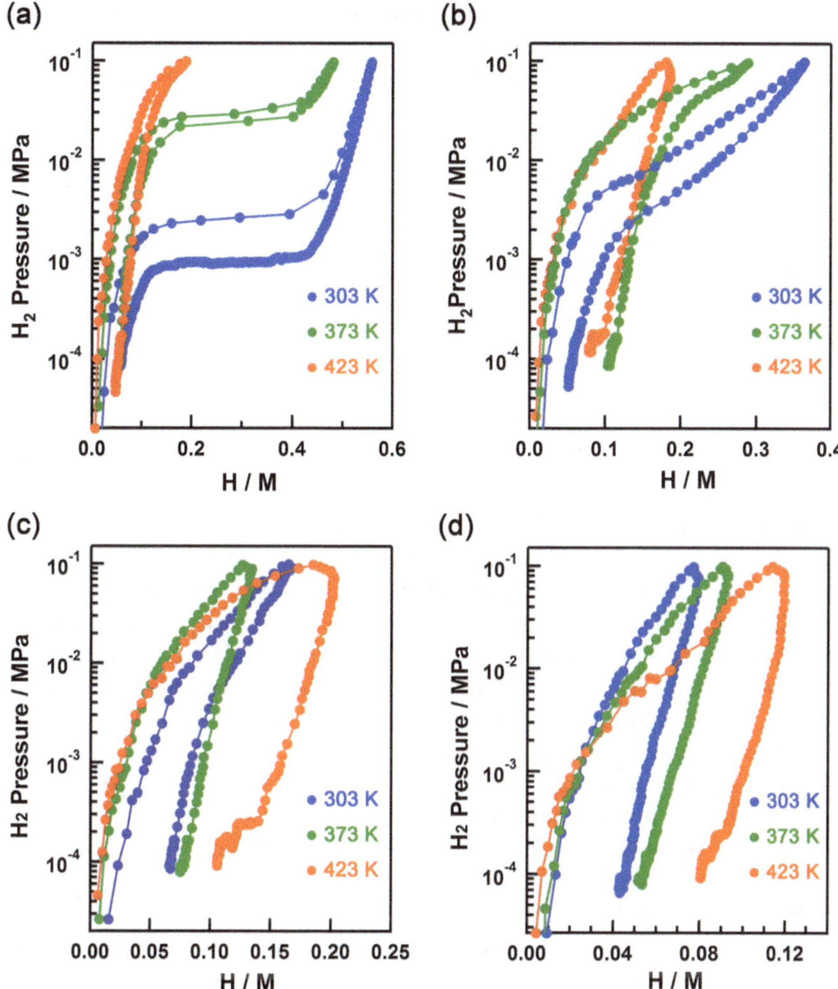

Fig. 3.18 The temperature dependences of PC isotherms in **a** Pd, **b** $Pd_{0.9}Ru_{0.1}$, **c** $Pd_{0.7}Ru_{0.3}$, **d** $Pd_{0.5}Ru_{0.5}$ nanoparticles

nanoparticles remain in α phase under a hydrogen pressure of 0.1 MPa. When the phase transition was treated as a chemical reaction, the author obtains

$$\frac{2}{C_\beta - C_\alpha}MH_{C_\alpha} + H_2 = \frac{2}{C_\beta - C_\alpha}MH_{C_\beta} + \Delta H_{\alpha\to\beta}$$

where $\Delta H_{\alpha\to\beta}$ is the heat of formation, C_α is the maximum hydrogen concentration in the α phase and C_β is the minimum hydrogen concentration in the β phase [10, 11]. C_α and C_β correspond to the start and end of the plateau region, respectively. As shown in Fig. 3.16, plateau-like region were observed in both Pd and $Pd_{0.9}Ru_{0.1}$ nanoparticles. An equilibrium pressure in this region is correlated with

Table 3.3 The heat of formation ($\Delta H_{\alpha \to \beta}$) and standard entropy ($\Delta S_{\alpha \to \beta}/R$) of hydride phase of Pd bulk, Pd nanoparticles and $Pd_{0.9}Ru_{0.1}$ nanoparticles

Sample	$\Delta H_{\alpha \to \beta}$ (kJ/mol H_2)	$\Delta S_{\alpha \to \beta}/R$ (mol $H_2)^{-1}$
Pd (Bulk) [41]	−40	−10
Pd nanoparticles	−32.81	−9.34
$Pd_{0.9}Ru_{0.1}$ nanoparticles	−18.74	−4.91

the $\Delta H_{\alpha \to \beta}$ and the entropy change ($\Delta S_{\alpha \to \beta}$) for the phase transition from the α phase to the β phase. In the case where one mole of hydrogen gas is consumed in the reaction, the thermodynamics are described below [41],

$$\ln \frac{P}{P_0} = \frac{\Delta H_{\alpha \to \beta}}{RT} - \frac{\Delta S_{\alpha \to \beta}}{R} \tag{3.1}$$

where P is the equilibrium hydrogen pressure and P_0 is the standard hydrogen pressure (0.1 MPa). From the Eq. (3.1), $\Delta H_{\alpha \to \beta}$ and $\Delta S_{\alpha \to \beta}$ were calculated and the results are summarized in Table 3.3. Both $\Delta H_{\alpha \to \beta}$ and $\Delta S_{\alpha \to \beta}$ for Pd nanoparticles are larger than those of Pd bulk. Furthermore those thermodynamics parameter increases with increasing Ru content. The $\Delta H_{\alpha \to \beta}$ is correlated with the bond strength between the metal and hydrogen in the β phase. As previous report has shown [42], this result also shows that the Pd nanoparticles have a larger $\Delta H_{\alpha \to \beta}$ value than Pd bulk. This implies that the bond strength between Pd and H atoms becomes weaker with decrease in particle size. On the other hand, the $\Delta H_{\alpha \to \beta}$ value is remarkably increased in $Pd_{0.9}Ru_{0.1}$ nanoparticles. The more the Ru content increases, the weaker the bond strength of M–H in the alloy. This phenomenon suggests a change in the electronic states of metals because of the formation of a solid solution alloy. The $\Delta S_{\alpha \to \beta}$ during the hydride formation comes mainly from the entropy loss of the hydrogen gas. The larger $\Delta S_{\alpha \to \beta}$ in $Pd_{0.9}Ru_{0.1}$ nanoparticles indicates that the hydrogen atoms possess a larger entropy compared with that in Pd nanoparticles. In other words, hydrogen atoms in the nanoparticles retain a part of the freedom in the gaseous state, implying that atomic arrangements or lattice defects in nanoparticles are statically or dynamically disordered because of the formation of a solid solution structure. From these results, the author succeeded in controlling the hydrogen absorption properties and electronic structure of Pd_xRu_{1-x} nanoparticles by metal composition over the whole composition range.

Solid-state ^2H NMR measurements were performed to investigate the state of ^2H in the Pd_xRu_{1-x} nanoparticles (Fig. 3.19). In the spectrum of Pd nanoparticles, a broad absorption line at 31 ppm and a sharp line around 0 ppm were observed. In the spectrum of $^2H_2$ gas, only a sharp line around 0 ppm was obtained. On comparison of these spectra, we can attribute the sharp line observed in the Pd nanoparticles to free deuterium gas ($^2H_2$) and the broad component to absorbed deuterium atoms (^2H) in the particles. Deuterium absorbed inside the lattice of Pd_xRu_{1-x} alloy nanoparticles also produced broad signals. Chemical shift values of the broad lines observed in Pd_xRu_{1-x} are summarized in Fig. 3.19b. Pd hydride shows the lowest-field shift of the ^2H atoms. The d band in Pd hydride is almost

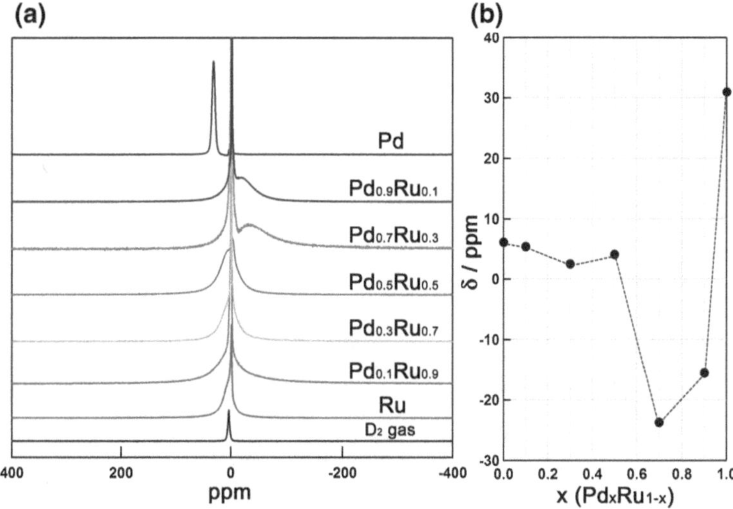

Fig. 3.19 **a** The solid-state ^2H NMR spectra for Pd_xRu_{1-x} nanoparticles and $^2H_2$ gas. All of the samples were measured under 101.3 kPa of $^2H_2$ gas at 303 K. **b** The chemical shift position of the broad absorption lines in Pd_xRu_{1-x} nanoparticles

filled [5], and so the d spin correlation is weakened, and the Knight shift is the dominant factor in the NMR shift of ^2H in Pd hydride. In contrast, for the Pd content of 70 and 90 at.%, the observed lines are markedly shifted upfield. This shift can be explained by the ^2H s electron's being polarized by the d spin paramagnetism of the PdRu nanoparticles arising from the formation of a hydride (Pd–Ru–H). Because a large number of holes exist in the d band of Pd–Ru, the spin correlation between the conduction electrons of the Pd–Ru hydride is large enough to cause spin polarization in the ^2H s electrons in the Pd–Ru hydride. The marked decrease in the shift value upon adding Ru to Pd from $x = 1.0$ to 0.7 can be explained by the increase of the effect from the paramagnetic spins in the d band.

With the further addition of Ru atoms (under 50 at.%), the shift value of the broad absorption lines, originating from ^2H inside the Pd–Ru lattice, jumped to near 0 ppm and became almost independent of the metal composition. This sudden change observed at $x = 0.5$ corresponds to the changeover of thermodynamic behavior from exothermic to endothermic and destabilization of the hydride phase that were observed in the PC isotherms. From the present result of NMR shift values, we conclude that remarkable changes in the electronic structure of Pd_xRu_{1-x} take place at $x = 0.5$ and between $x = 0.9$ and 1.0.

To investigate the CO-oxidizing catalytic activity, Pd_xRu_{1-x} alloy, Rh nanoparticles and a physical mixture (Ru and Pd nanoparticles) supported on γ-Al_2O_3 catalysts were prepared by wet impregnation method. The catalysts were heated in increments of 10 °C to a temperature at which CO was consumed completely, and the products were analyzed at each temperature. Figure 3.20 compares the CO conversions for 1 wt% of each nanoparticle supported on γ-Al_2O_3.

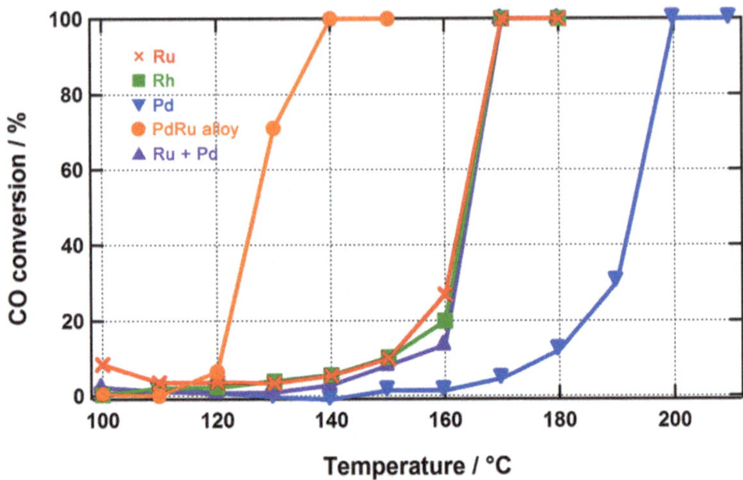

Fig. 3.20 Temperature dependence of CO conversion in Ru (*cross red*), Rh (*filled square green*), Pd (*filled down pointed triangle blue*), $Pd_{0.5}Ru_{0.5}$ solid solution (*filled circle orange*) and Ru + Pd mixture (*filled up pointed triangle purple*) nanoparticles supported on γ-Al_2O_3

The temperatures corresponding to 50 % conversion (T_{50}) of CO to CO_2 for $Pd_{0.5}Ru_{0.5}$, Ru, Rh and Pd catalysts were approximately 125, 165, 165 and 195 °C, respectively. The temperatures for 100 % conversion (T_{100}) of CO to CO_2 were 140, 170, 170 and 200 °C, respectively. From these results, it is clear that the $Pd_{0.5}Ru_{0.5}$ nanoparticles have the highest activity for 50 and 100 % conversions of CO to CO_2. $Pd_{0.5}Ru_{0.5}$ shows a 30–40 K lower conversion temperature compared with the most efficient CO-oxidation catalyst, Ru. In addition, the conversion temperature was lower than that for the most expensive precious metal, Rh. That is, $Pd_{0.5}Ru_{0.5}$ nanoparticles are more efficient CO-oxidation catalyst than Rh. On the other hand, the conversion property of the physical mixture of Ru and Pd nanoparticles differs totally from that of $Pd_{0.5}Ru_{0.5}$. The physical mixture of Ru and Pd nanoparticles shows the same conversion property as that of Ru nanoparticles because the Ru nanoparticle is better and dominant in the CO-oxidation reaction compared with the Pd nanoparticle. These results demonstrate a novel strategy to create functional materials on the basis of inter-elemental fusion. The extreme enhancement of the CO-oxidizing ability observed for $Pd_{0.5}Ru_{0.5}$ alloy is considered to originate from a new electronic state realized by the atomic-level alloying of Pd and Ru.

In order to further investigate the catalytic activity of Pd_xRu_{1-x} nanoparticles, the temperature dependence of CO conversion for each metal composition nanoparticle was measured. Figure 3.21 shows the metal composition dependence of the CO conversion in Pd_xRu_{1-x} nanoparticles supported on γ-Al_2O_3. The T_{50} and T_{100} in all of the alloy nonoparticles are lower than those in pure Ru or Pd nanoparticles, and those temperatures reached a minimum at the ratio of Pd:Ru = 50:50. From these results, it is revealed that the $Pd_{0.5}Ru_{0.5}$ nanoparticles

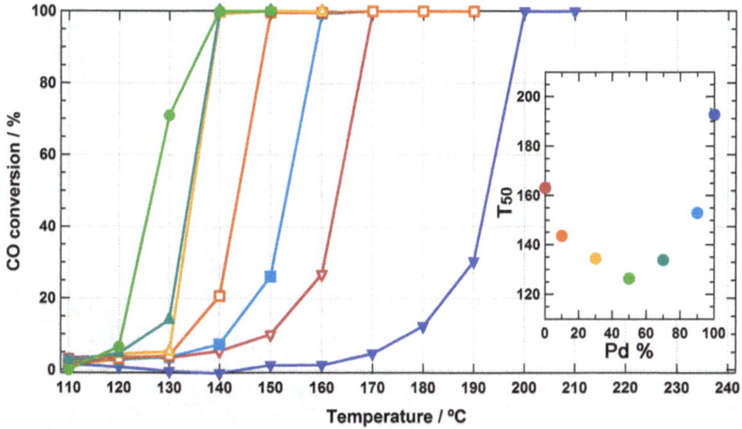

Fig. 3.21 Temperature dependence of CO conversion in Pd_xRu_{1-x} nanoparticles supported on γ-Al_2O_3; x = 0 (*down pointed triangle red*), 0.1 (*square orange*), 0.3 (*up pointed triangle yellow*), 0.5 (*filled circle green*), 0.7 (*filled up pointed triangle blue-green*), 0.9 (*filled square light blue*), and 1.0 (*filled down pointed triangle blue*). The inset is the metal composition dependence of T_{50}

have the highest activity for 50 and 100 % conversions of CO to CO_2. This kind of inverse volcano type behavior for CO oxidation has been reported in many types of alloy nanoparticle catalysts such as Au–Pd [43] and Au–Ag [44, 45]. It has been reported that the CO-oxidation reaction takes place between chemisorbed species (Langmuir–Hinshelwood mechanism), and thus O + CO co-adsorption was considered as the initial states of the oxidation reaction [46–48]. The position of the d-band center, relative to the Fermi level is an important parameter controlling adsorption of O and CO or activation energy barriers for CO oxidation [49, 50]. By the atomic-level alloying, the d-band center of $Pd_{0.5}Ru_{0.5}$ is considered to change to a favorable condition for CO and O coverage and/or high reactivity of CO with O atoms on the bimetallic surface. In addition, it has been reported that defect sites can also significantly affect the catalytic activity [43]. The coexistence of both fcc and hcp domains in a Pd–Ru single particle is considered to generate defect structures with vacancies in the domain boundaries, leading to the high reactivity of CO with O atoms. A theoretical calculation to clarify the mechanism and the correlation between metal composition and the CO conversion for Pd_xRu_{1-x} nanoparticles is currently in progress.

To confirm that the solid solution structure was maintained in the nanoparticles before and after the catalytic reaction, the HRSTEM, high-angle annular dark-field (HAADF) STEM and EDX analyses were measured by JEM-ARM200F operated at 200 kV accelerate voltage. Figure 3.22 shows elemental mapping data for the $Pd_{0.5}Ru_{0.5}$ catalyst after the CO-oxidizing reaction. Figure 3.22a and b are a BFSTEM image and HAADF STEM image, respectively. Figures 3.22c, d and e are the corresponding Pd–L, Ru–L and Al–K STEM–EDX maps, respectively. Figure 3.22f is an overlay map of the Pd, Ru and Al chemical distributions. As

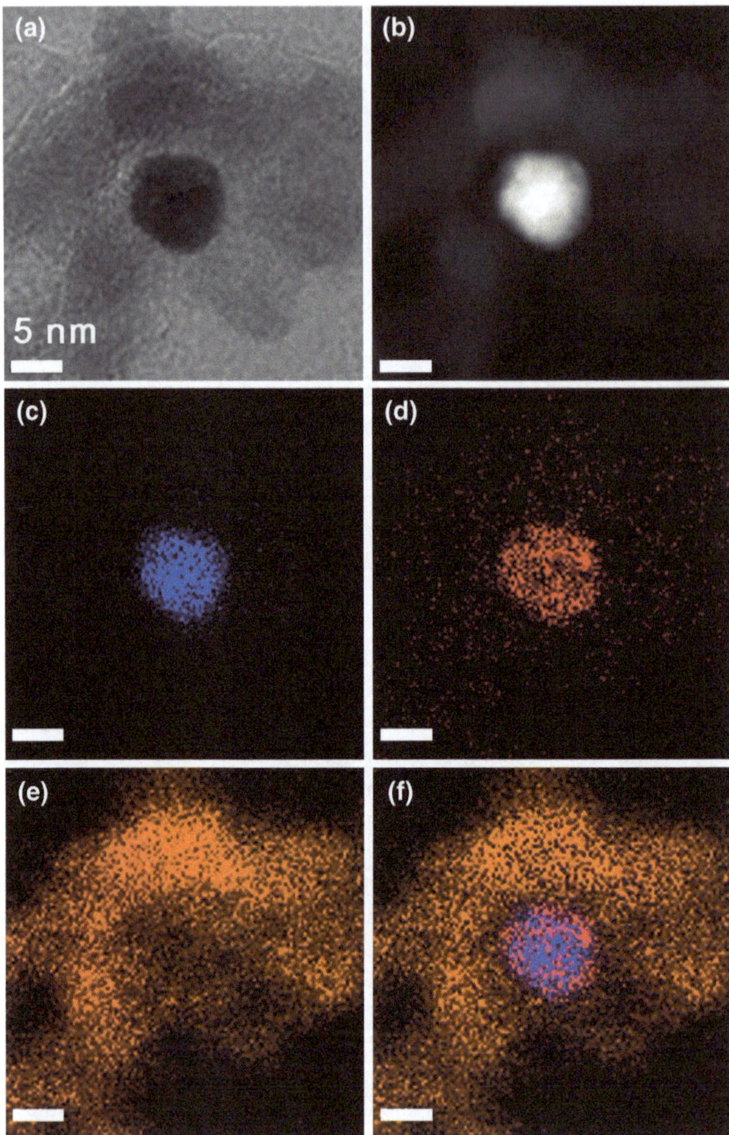

Fig. 3.22 **a** BF STEM image, **b** HAADF–STEM image, **c** Pd–L STEM–EDX map, **d** Ru–L STEM–EDX map and **e** Al–K STEM–EDX map obtained for 1 wt% of $Pd_{0.5}Ru_{0.5}$ nanoparticle supported on γ-Al_2O_3 after measuring CO-oxidizing catalytic activity. **f** Reconstructed overlay image of the maps shown in panels **c**, **d** and **e** (*blue* Pd; *red* Ru; *orange* Al)

Fig. 3.22f shows, a nanoparticle is supported on γ-Al_2O_3 and the nanoparticle maintain the solid solution structure after the reaction. The novel PdRu nanostructured materials have not only high catalytic activity but also high durability for the CO-oxidation reaction.

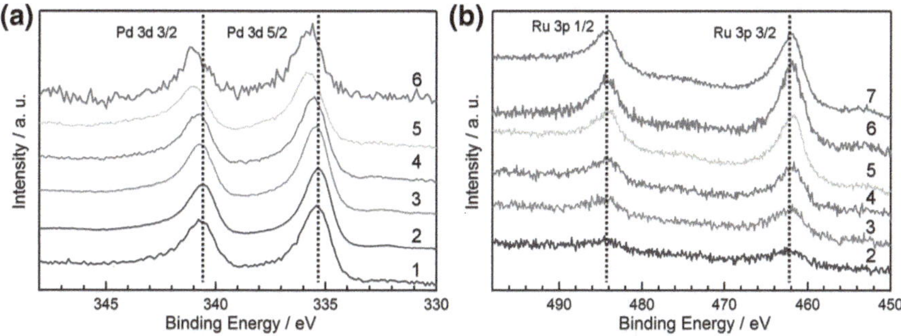

Fig. 3.23 The (**a**) Pd 3d and (**b**) Ru 3p core-level XPS of Pd_xRu_{1-x} nanoparticles (1:Pd, 2:$Pd_{0.9}Ru_{0.1}$, 3:$Pd_{0.7}Ru_{0.3}$, 4:$Pd_{0.5}Ru_{0.5}$, 5:$Pd_{0.3}Ru_{0.7}$, 6:$Pd_{0.1}Ru_{0.9}$ and 7: Ru nanoparticles)

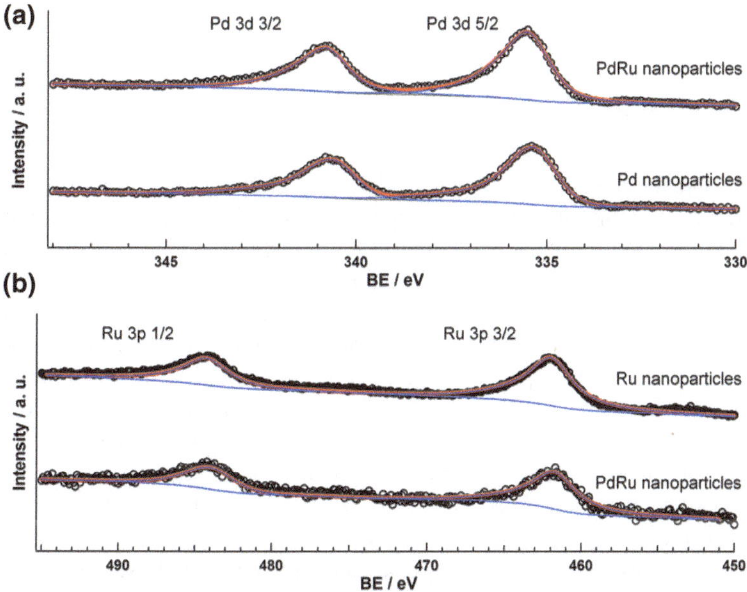

Fig. 3.24 (**a**) Pd 3d and (**b**) Ru 3p core-level XPS of Pd, $Pd_{0.5}Ru_{0.5}$ and Ru nanoparticles

The change in the electronic structure because of atomic-level alloying of Ru and Pd was investigated by XPS measurement (Figs. 3.23, 3.24). XPS spectra for samples on a carbon sheet were recorded using a Shimadzu ESCA-3400 X-ray photoelectron spectrometer. The binding energies were corrected with reference to the C(1s) line at 284.5 eV. Figures 3.23, 3.24 show the Pd 3d and Ru 3p core-level XPS spectra of samples. The Pd 3d and Ru 3p binding energies are summarized in Table 3.4. The Pd 3d binding energies of the alloys shifted to higher energy with increasing Ru content, which is in agreement with the oxidization of Pd. By contrast, the Ru 3p binding energies of alloys shifted to lower energy with increasing

Table 3.4 Binding energies of Pd 3d and Ru 3p

Sample	BE(Pd $3d_{3/2}$) (eV)	BE(Pd $3d_{5/2}$) (eV)	BE(Ru $3p_{1/2}$) (eV)	BE(Ru $3p_{3/2}$) (eV)
Pd	340.65	335.38		
$Pd_{0.5}Ru_{0.5}$	340.80	335.51	484.00	461.75
Ru			484.24	461.99

Pd content, which is in agreement with the reduction of Ru. Thus, the XPS spectra for Pd_xRu_{1-x} nanoparticles indicate that the electronic state of Pd_xRu_{1-x} nanoparticles differs from those of Ru and Pd nanoparticles and that electrons of Pd atoms slightly transfer to Ru atoms. This electronic transfer may also be responsible for the catalytic enhancement, which would be consistent with d-band theory suggested in previous reports [50, 51].

3.4 Conclusion

In summary, the author first obtained Pd_xRu_{1-x} solid solution alloy nanoparticles over the whole composition range through the chemical reduction method, although Ru and Pd are immiscible at the atomic level in the bulk state. From the XRD and TEM analyses, it was found that the structure of Pd_xRu_{1-x} changes from fcc to hcp with increasing Ru content. The structures of Pd_xRu_{1-x} nanoparticles in the Pd composition ranges of 30–70 % consisted of both fcc and hcp structures, and both phases coexist in a single particle. In addition, the amount of hydrogen absorption depends on the metal composition of the alloy; the hydrogen capacities of Pd_xRu_{1-x} nanoparticles are decreased with decreasing Pd content. And The reaction of hydrogen with the Pd_xRu_{1-x} nanoparticles changed from exothermic to endothermic as the Ru content increased. Accordingly, this difference in the amounts of hydrogen absorption implies a difference in the electronic structures of the Pd_xRu_{1-x} nanoparticles. Furthermore, the prepared PdRu nanoparticles exhibit extremely enhanced

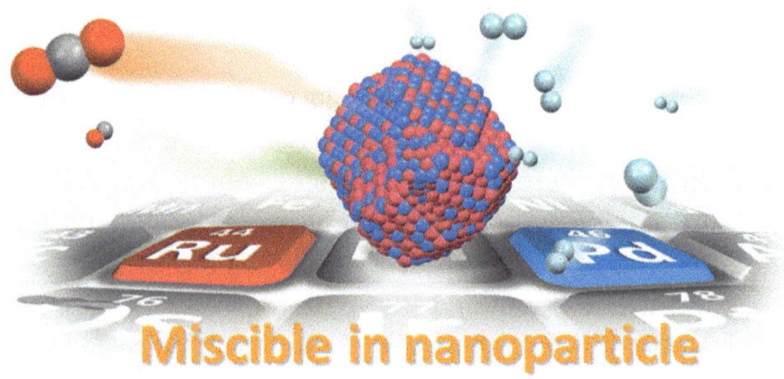

Fig. 3.25 Graphical abstraction of Pd–Ru

CO-oxidizing catalytic activity; $Pd_{0.5}Ru_{0.5}$ nanoparticles showed the highest catalytic activity, and this activity is also higher than that of Rh. In addition, from the point of view of elemental strategy [33], the 1:1 PdRu solid solution alloy is expected to act as a more efficient catalyst than Rh. Although Pd and Ru are cheaper than Rh, the catalytic activity of $Pd_{0.5}Ru_{0.5}$ is higher than that of Rh. This work provides a novel strategy on the basis of inter-elemental fusion to create highly efficient functional materials for energy and material conversions (Fig. 3.25).

References

1. Tripathi SN, Bharadwaj SR, Chandransekharaiah MS (1996) The Rh–Ru system (rhodium–ruthenium). J Phase Equil 17:362–365
2. Massalski TB, Okamoto H, Subramanian PR, Kacprzak L (1996) Binary alloy phase diagrams. ASM International, Materials Park
3. Tripathi SN, Bharadwaj SR, Dharwadkar SR (1993) The Pd–Ru system (palladium–ruthenium). J Phase Equil 14:638–642
4. Kusada K, Yamauchi M, Kobayashi H, Kitagawa H, Kubota Y (2010) Hydrogen-storage properties of solid-solution alloys of immiscible neighboring elements with Pd. J Am Chem Soc 132:15896–15898
5. Papaconstantopoulos DA, Klein BM, Economou EN, Boyer LL (1978) Band structure and superconductivity of PdD_x and PdH_x. Phys Rev B 17:141–150
6. Gelatt CD, Ehren-reich H, Weiss JA (1978) Transition-metal hydrides: electronic structure and the heats of formation. Phys Rev B 17:1940–1957
7. Vuillemin JJ, Priestly MG (1965) De Haas-Van Alphen effect and fermi surface in palladium. Phys Rev Lett 14:307–308
8. Mueller FM, Freeman AJ, Dimmock JO, Furdyna AM (1970) Electronic structure of palladium. Phys Rev B 1:4617–4634
9. Wicke E (1984) Electronic structure and properties of hydrides of 3D and 4D metals and intermetallics. J Less Common Met 101:17–33
10. Alefeld G, Völkl J (1978) Hydrogen in metals I. Springer, Berlin, p 108
11. Alefeld G, Völkl J (1978) Hydrogen in metals II. Springer, Berlin, p 73
12. Ke X, Kramer GJ, Løvvik OM (2004) The influence of electronic structure on hydrogen absorption in palladium alloys. J Phys Condens Matter 16:6267–6278
13. Flanagan TB, Wang D, Noh H (1997) The effect of cycling through the hydride phase on isotherms for fcc Pd-rich alloys. J Alloys Compd 253–254:216–220
14. Barlag H, Opara L, Züchner H (2002) Hydrogen diffusion in palladium based fcc alloys. J Alloys Compd 330–332:434–437
15. Sonwane CG, Wilcox J, Ma YH (2006) Solubility of hydrogen in PdAg and PdAu binary alloys using density functional theory. J Phys Chem B 110:24549–24558
16. Cabrera AL, Morales EL, Hansen J, Schuller K (1995) Structural changes induced by hydrogen absorption in palladium and palladium-ruthenium alloys. Appl Phys Lett 66:1216–1218
17. Frölich K, Severin HG, Hempelmann R, Wicke E (1980) Local magnetic moments of ruthenium in palladium/ruthenium/hydrogen alloys. Z Phys Chem Neue Fol 119:33–52
18. Szafranski AW (2003) Influence of hydrogen on the thermoelectronic power of palladium alloyed with neighbouring elements: I. Pd/Ru/H and Pd/Rh/H alloys. J Phys Condens Matter 15:3583–3592
19. Perkas N, Teo J, Shen S, Wang Z, Highfield J, Zhong Z, Gedanken A (2011) Supported Ru catalysts prepared by two sonication-assisted methods for preferential oxidation of CO in H_2. Phys Chem Chem Phys 13:15690–15698

20. Bowker M (2007) Automotive catalysis studied by surface science. Chem Soc Rev 37:2204–2211
21. Grass ME, Zhang Y, Butcher DR, Park JY, Li Y, Bluhm H, Bratlie KM, Zhang T, Somorjai GA (2008) A reactive oxide overlayer on rhodium nanoparticles during CO oxidation and its size dependence studied by in situ ambient-pressure X-ray photoelectron spectroscopy. Angew Chem Int Ed 47:8893–8896
22. Zhang Y, Grass ME, Huang W, Somorjai GA (2010) Seedless polyol synthesis and CO oxidation activity of monodisperse (111)- and (100)-oriented rhodium nanocrystals in sub-10 nm sizes. Langmuir 26:16463–16468
23. Joo SH, Park JK, Renzas JR, Butcher DR, Huang W, Somorjai GA (2010) Size effect of ruthenium nanoparticles in catalytic carbon monoxide oxidation. Nano Lett 10:2709–2713
24. Ertl G (2008) Reactions at surfaces: from atoms to complexity (Nobel lecture). Angew Chem Int Ed 47:3524–3535
25. Xiao L, Zhuang L, Liu Y, Lu J, Abruna HD (2009) Activating Pd by morphology tailoring for oxygen reduction. J Am Chem Soc 131:602–608
26. Newton MA, Belver-Coldeira C, Martínez-Arias A, Fernández-García M (2007) Dynamic in situ observation of rapid size and shape change of supported Pd nanoparticles during CO/NO cycling. Nat Mater 6:528–532
27. Schultz NE, Gherman BE, Cramer CJ, Truhlar DG (2006) Pd_nCO (n = 1,2): accurate Ab initio bond energies, geometries, and dipole moments and the applicability of density functional theory for fuel cell modeling. J Phys Chem B 110:24030–24046
28. Lee HC, Potapova Y, Lee D (2012) A core-shell structured, metal–ceramic composite-supported Ru catalyst for methane steam reforming. J Power Sources 216:256–260
29. McFarland E (2012) Unconventional chemistry for unconventional natural gas. Science 338:340–342
30. Beutl M, Lesnik J (2001) Adsorption dynamics of hydrogen and carbon monoxide on V/Pd(111) alloy surface. Surf Sci 482–485:353–358
31. Abdelsayed V, Aljarash A, El-Shall MS, Othman ZAA, Alghamdi AH (2009) Microwave synthesis of bimetallic nanoalloys and CO oxidation on ceria-supported nanoalloys. Chem Mater 21:2825–2834
32. Renzas JR, Huang W, Zhang Y, Grass ME, Hoang DT, Alayoglu S, Butcher DR, Tao F, Liu Z, Somorjai GA (2011) $Rh_{1-x}Pd_x$ nanoparticle composition dependence in CO oxidation by oxygen: catalytic activity enhancement in bimetallic systems. Phys Chem Chem Phys 13:2556–2562
33. Nakamura E, Sato K (2011) Managing the scarcity of chemical elements. Nat Mater 10:158–161
34. Kobayashi H, Yamauchi M, Kitagawa H, Kubota Y, Kato K, Takata M (2010) Atomic-level Pd-Pt alloying and largely enhanced hydrogen-storage capacity in bimetallic nanoparticles reconstructed from core/shell structure by a process of hydrogen absorption/desorption. J Am Chem Soc 132:5576–5577
35. Kakade BA, Tamaki T, Ohashi H, Yamaguchi T (2012) Highly active bimetallic PdPt and CoPt nanocrystals for methanol electro-oxidation. J Phys Chem C 116:7464–7470
36. Kobayashi H, Morita H, Yamauchi M, Ikeda R, Kitagawa H, Kubota Y, Kato M, Takata M, Toh S, Matsumura S (2012) Nanosize-induced drastic drop in equilibrium hydrogen pressure for hydride formation and structural stabilization in Pd-Rh solid-solution alloys. J Am Chem Soc 134:12390–12394
37. Petkov V, Wanjala BN, Loukrakpam R, Luo J, Yang L, Zhong C, Shastri S (2012) Pt–Au alloying at the nanoscale. Nano Lett 12:4289–4299
38. Hernández-Fernández P, Rojas S, Ocón P, Gómez de la Fuente JL, San Fabián J, Sanza J, Peña MA, García-García FJ, Terreros P, Fierro JLG (2007) Influence of the preparation route of bimetallic Pt–Au nanoparticle electrocatalysts for the oxygen reduction reaction. J Phys Chem C 111:2913–2923

39. Essinger-Hileman ER, DeCicco D, Bondi JF, Schaak RE (2011) Aqueous room-temperature synthesis of Au–Rh, Au–Pt, Pt–Rh, and Pd–Rh alloy nanoparticles: fully tunable compositions within the miscibility gaps. J Mater Chem 21:11599–11604
40. Denton AR, Ashcroft NW (1991) Vegard's law. Phys Rev A 43:3161–3164
41. Fukai Y (2005) The metal-hydrogen system, basic bulk properties, 2nd edn. Springer, Berlin
42. Yamauchi M, Ikeda R, Kitagawa H, Takata M (2008) Nanosize effects on hydrogen storage in palladium. J Phys Chem C 112:3294–3299
43. Xu J, White T, Li P, He C, Yu J, Yuan W, Han Y (2010) Biphasic Pd-Au alloy catalyst for low-temperature CO oxidation. J Am Chem Soc 132:10398–10406
44. Sandoval A, Aguilar A, Louis C, Traverse A, Zanella R (2011) Bimetallic Au–Ag/TiO_2 catalyst prepared by deposition–precipitation: high activity and stability in CO oxidation. J Catal 281:40–49
45. Wang A, Liu J, Lin SD, Lin T, Mou C (2005) A novel efficient Au–Ag alloy catalyst system: preparation, activity, and characterization. J Catal 233:186–197
46. Engel T, Ertl G (1978) A molecular beam investigation of the catalytic oxidation of CO on Pd (111). J Chem Phys 69:1267–1281
47. Ladas S, Poppa H, Boudart M (1981) The adsorption and catalytic oxidation of carbon monoxide on evaporated palladium particles. Surf Sci 102:151–171
48. Liu K, Wang A, Zhang T (2012) Recent advances in preferential oxidation of CO reaction over platinum group metal catalysts. ACS Catal 2:1165–1178
49. Hammer B, Nørskov JK (1995) Why gold is the noblest of all the metals. Nature 376:238–240
50. Hammer B, Nørskov JK (1995) Electronic factors determining the reactivity of metal surfaces. Surf Sci 343:211–220
51. Nikolla E, Schwank J, Linic S (2009) Measuring and relating the electronic structures of nonmodel supported catalytic materials to their performance. J Am Chem Soc 131:2747–2754

Chapter 4
Discovery of the Face-Centered Cubic Ruthenium Nanoparticles: Facile Size-Controlled Synthesis Using the Chemical Reduction Method

4.1 Introduction

The majority of metals have one of the three basic structures: body-centered cubic (bcc), hexagonal close-packed (hcp) or face-centered cubic (fcc). In the periodic table, the canonical hcp–bcc–hcp–fcc structural sequence is well known as the atomic number increases across the transition metal series, and the relative stability of the structures is determined by the total electronic energy of the metals [1]. The study of pressure–temperature (P–T) phase diagrams for the bulk metals have been extensively investigated, [2–11] but when the size is reduced to nano-dimensions it has been reported that the phase diagrams for metals change considerably from the bulk [12–14]. For example, fcc Co and fcc Fe nanoparticles are stabilized at ambient conditions, even though these phases only exist at high temperature in bulk [15–17].

Recently Ru, a 4d transition metal which adopts a hcp structure at all temperature ranges in bulk, has attracted much attention as a CO oxidizing catalyst due to its high catalytic activity [18–22]. CO oxidation catalysts have been extensively investigated recently because of their potential application for CO removal from car exhaust or for preventing CO poisoning in fuel cell systems [23–30]. In this chapter, the author reports the first example of fcc-structured Ru and also demonstrates a facile synthesis method to control the size of the nanoparticles. The structure and size-dependent catalytic activity of CO oxidation was observed over Ru nanoparticles.

K. Kusada, *Creation of New Metal Nanoparticles and Their Hydrogen-Storage and Catalytic Properties*, Springer Theses, DOI 10.1007/978-4-431-55087-7_4

Table 4.1 Reaction conditions for the syntheses of fcc and hcp Ru nanoparticles

Sample	Structure	Size (nm)	Metal precursor/(mmol)	Solvent/(ml)	PVP (mmol)
A	Fcc	2.4 ± 0.5	$Ru(acac)_3$/2.1	TEG/500	10.0
B	Fcc	3.5 ± 0.7	$Ru(acac)_3$/2.1	TEG/200	10.0
C	Fcc	3.9 ± 0.8	$Ru(acac)_3$/2.1	TEG/100	5.0
D	Fcc	5.4 ± 1.1	$Ru(acac)_3$/2.1	TEG/25	1.0
E	Hcp	2.2 ± 0.5	$RuCl_3 \cdot nH_2O$/2.1	EG/500	10.0
F	Hcp	3.5 ± 0.6	$RuCl_3 \cdot nH_2O$/2.1	EG/200	10.0
G	Hcp	3.9 ± 0.6	$RuCl_3 \cdot nH_2O$/2.1	EG/100	5.0
H	Hcp	5.0 ± 0.7	$RuCl_3 \cdot nH_2O$/2.1	EG/25	1.0

4.2 Experiment

4.2.1 Syntheses of Ru Nanoparticles

Uniformly-sized fcc and hcp Ru nanoparticles with sizes ranging from 2 to 5.5 nm were prepared by chemical reduction methods using $RuCl_3 \cdot nH_2O$ and Ruthenium(III) acetylacetonate ($Ru(acac)_3$) as the metal precursors and poly (*N*-vinyl-2-pyrrolidone) (PVP) as the stabilizing agent. Ethylene glycol (EG) and triethylene glycol (TEG) acted as solvents and the reducing agents for the synthesis. Phase control was achieved by varying both the type of metal precursor and solvent, and size control was achieved through adjusting the concentration of reagents and the PVP stabilizer used for the synthesis (Table 4.1). In a typical synthesis of fcc Ru nanoparticles having a diameter of 2.4 nm, $Ru(acac)_3$ (2.1 mmol) and PVP (10 mmol, in terms of monomer unit) were dissolved in TEG (500 ml) at room temperature. The solution was then heated to 200 °C and maintained at this temperature for 3 h. After the reaction was complete, the prepared nanoparticles were separated by centrifugation. The syntheses of fcc and hcp Ru nanoparticles larger than 3.5 nm were performed based on this method, but an adequate amount of PVP was added into the reaction solution after 3 h heating to adjust the relative amount of PVP to be similar to the smaller sized nanoparticles.

4.2.2 Catalysts Preparation

To investigate the CO-oxidizing catalytic activity both fcc and hcp Ru nanoparticles supported on γ-Al_2O_3 catalysts were prepared by wet impregnation. Each nanoparticle (equivalent to 1 wt% of γ-Al_2O_3) was ultrasonically dispersed in purified water. γ-Al_2O_3 support that had been precalcined at 1,073 K for 5 h was put in each aqueous solution of nanoparticles and the suspended solutions were stirred for 12 h. After stirring, the suspended solutions were heated to 60 °C and dried under vacuum. The resulting powders were kept at 120 °C for 8 h to remove water completely.

4.2.3 *Characterizations*

Transmission Electron Microscopy Analysis

TEM images were recorded on a Hitachi HT 7700 TEM instrument operated at 100 kV accelerate voltage. The samples dispersed with ethanol were drop-cast onto a carbon-coated copper grid and allowed to dry under ambient conditions.

High Resolution TEM Analysis

HRTEM images were captured using a JEOL ARM 200F STEM instrument operated at 200 kV. The samples dispersed with ethanol were drop-cast onto a carbon-coated copper grid and allowed to dry under ambient conditions.

X-Ray Diffraction Measurement

The crystal structures of Ru nanoparticles were investigated by powder XRD analysis collected on a Bruker D8 Advance diffractometer (Cu K_{α} radiation).

In Situ XRD Measurement

The thermal stability of fcc Ru nanoparticles were investigated by in situ powder XRD analysis measured at the BL02B2 beamline, SPring-8. The XRD patterns of the samples sealed in a glass capillary were measured in situ under vacuum condition at temperature range between 303 K and 723 K. The wavelength is 0.578375(1) Å.

Catalytic Test

The obtained catalyst powders were pressed into pellets at 1.2 MPa for 5 min. The pellets were crushed and sieved to obtain grains with diameters between 180 and 250 μm. Each supported nanoparticle catalyst (150 mg) was loaded into a tubular quartz reactor (i.d. 7 mm) with quartz wool. $CO/O_2/He$ mixed gas ($He/CO/O_2$: 49/0.5/0.5 ml min^{-1}) was passed over the catalysts at ambient temperature, and the catalysts were then heated to 100 °C. After 15 min, effluent gas was collected and the reaction products were analyzed by gas chromatography with a thermal conductivity detector (GC-8A, Shimadzu, Japan). Catalysts were heated in increments of 10 °C to a temperature at which CO was consumed completely, and the products were analyzed at each temperature. After the reaction, the reactor was purged with He at the reaction temperature, and the catalysts were then cooled to room temperature.

4.3 Results and Discussion

TEM images for the synthesized samples were recorded on a Hitachi HT7700 TEM instrument operated at 100 kV. TEM images in Fig. 4.1a–h show the formation of uniform Ru nanoparticles having a diameter of 2.2–5.4 nm with narrow size distributions. The mean diameters and distributions were summarized in Table 4.1, which were estimated by averaging over at least 200 particles.

The crystal structures of Ru nanoparticles were investigated by powder X-ray diffraction (XRD) analysis collected on a Bruker D8 Advance diffractometer (Cu K_{α} radiation). Figure 4.2 shows the patterns of each prepared Ru nanoparticle. The trend in the XRD patterns correlates well with the TEM images, as the line widths of the diffraction peaks become sharper with increasing the particle mean diameter size. One of the most noteworthy outcomes in this study is that the structure of Ru nanoparticle is controllable only by choosing adequate combinations of a metal precursor and a reducing agent. All the Ru nanoparticles synthesized with $Ru(acac)_3$ and TEG adopt a fcc structure, while the Ru nanoparticles synthesized with $RuCl_3 \cdot nH_2O$ and EG have a hcp structure. While several calculation studies reported the possibility of fcc phase of Ru, [31, 32] to the best of the author's knowledge, this is the first report of a synthesis of fcc Ru.

The structures of prepared nanoparticles were also confirmed through HRTEM analysis. Typical examples of both fcc and hcp nanoparticles are shown in Fig. 4.3. Figure 4.3b is a hcp Ru nanoparticle recorded along the (100) direction. It clearly shows the ABABAB… stacking sequence of hcp. On the other hand, Fig. 4.3a is a Ru nanoparticle representing (111) fcc planes. The single particle is a five-fold symmetry twinned nanoparticle having decahedral structure, which consists of five tetrahedrons. The decahedral structure is well known as a typical structure of fcc

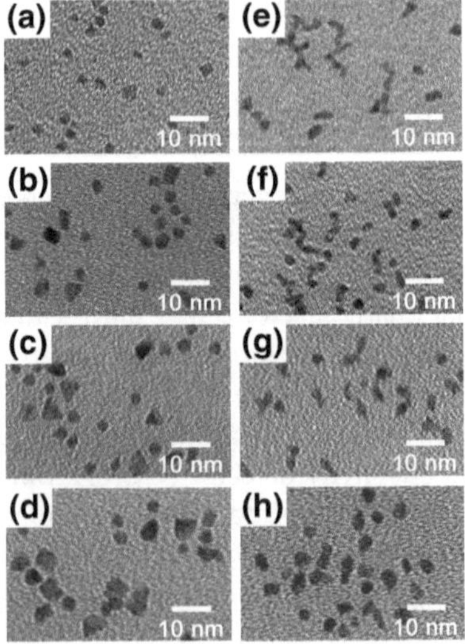

Fig. 4.1 The TEM images of synthesized Ru nanoparticles; **a** 2.4, **b** 3.5, **c** 3.9, **d** 5.4, **e** 2.2, **f** 3.5, **g** 3.9 and **h** 5.0 nm

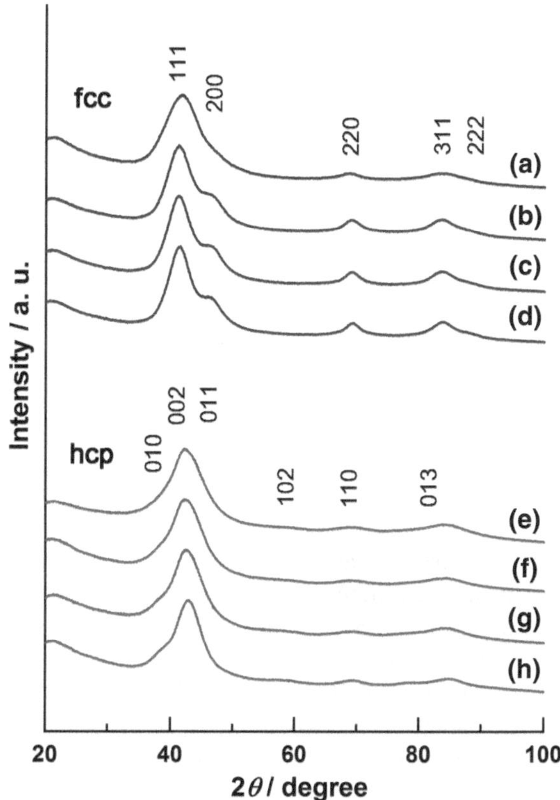

Fig. 4.2 The XRD patterns for face-centered cubic Ru nanoparticles (*A*–*D*) and hexagonal close-packed Ru nanoparticles (*E*–*H*) at room temperature. The radiation wavelength is 1.54 Å (Cu K_{α})

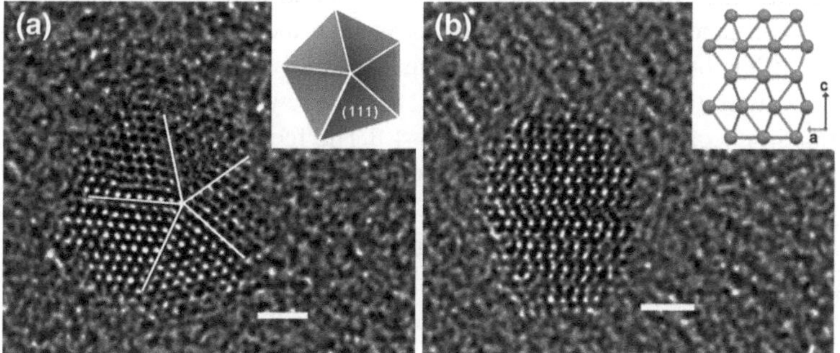

Fig. 4.3 **a** The HRTEM image of a fcc nanoparticle in sample *D*. The inset is an illustration of the decahedral structure. **b** The HRTEM image of a hcp nanoparticle in sample *H*. The inset is an illustration of hcp lattice viewed along the (100) direction. Scale bars are 1.0 nm

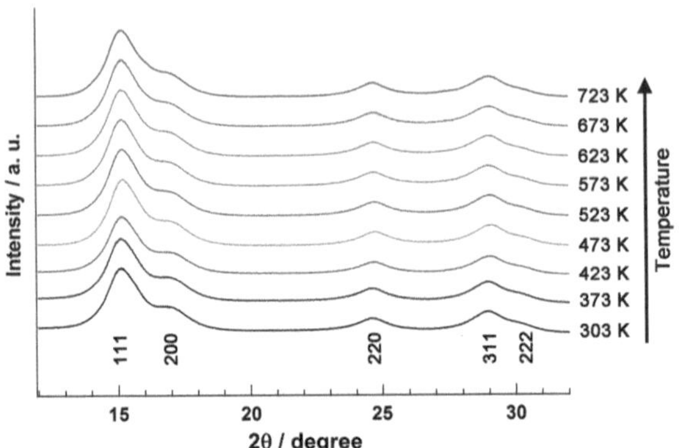

Fig. 4.4 The in situ XRD patterns of fcc Ru nanoparticles measured under vacuum condition at temperature range between 303 and 723 K. The wavelength is 0.578375(1) Å

metal nanoparticles, such as Au, Ag and Pd. [33, 34] From these results, it was confirmed that the crystal structure of Ru nanoparticles is controllable by adjusting the synthetic precursors.

To further investigate the thermal stability of fcc Ru nanoparticles, in situ powder XRD measurements were performed at the beamline BL02B2, at SPring-8 (Fig. 4.4). It was found that the prepared Ru nanoparticles show the fcc structure at a wide range of temperatures and are stable up to 723 K.

In order to investigate the catalytic activity of the synthesized nanoparticles for CO oxidation, the Ru nanoparticles were supported on γ-Al_2O_3 by a wet impregnation method. The catalysts were heated in increments of 10 °C to a temperature at which CO was consumed completely, and the products were analyzed at each temperature. The temperature dependencies of CO conversion in Ru nanoparticles supported on γ-Al_2O_3 are summarized in Fig. 4.5. Figure 4.6 compares the CO conversions for 1 wt% of each nanoparticle supported on γ-Al_2O_3. In hcp Ru nanoparticles, the temperature for 50 % conversion of CO to CO_2 (T_{50}) increased with increasing particle size, a trend which is similar to Au or Rh nanoparticles. [35, 36] Conversely, in fcc Ru nanoparticles, T_{50} decreased with increasing particle size, which is similar to the trend observed for Pt nanoparticles. [37] Above 3 nm, the novel fcc Ru nanoparticles are more reactive than the conventional hcp Ru nanoparticles. It has been reported that the mechanism for CO oxidation with hcp Ru begins first with the oxidation of Ru (001) into a few RuO_2 (110) layers, and the CO oxidation occurs on RuO_2 (110). [38–40] The fcc (111) is also a close-packed plane in common with the hcp (001). Although hcp nanoparticles are not completely enclosed with only close-packed planes, fcc nanoparticles generally tend to be enclosed by (111) planes because the fcc (111) planes have the lowest surface energy. [41] Therefore, fcc Ru surface could be more reactive than hcp Ru for CO oxidation. However the size dependences are different. This tendency might result from the differences in the formation process of oxide layers, the adsorption behavior of CO or the activation energy, due to the differences of electronic states or surfaces in fcc and hcp structures. Further studies including theoretical calculations are required to understand the oxidation mechanism for fcc Ru.

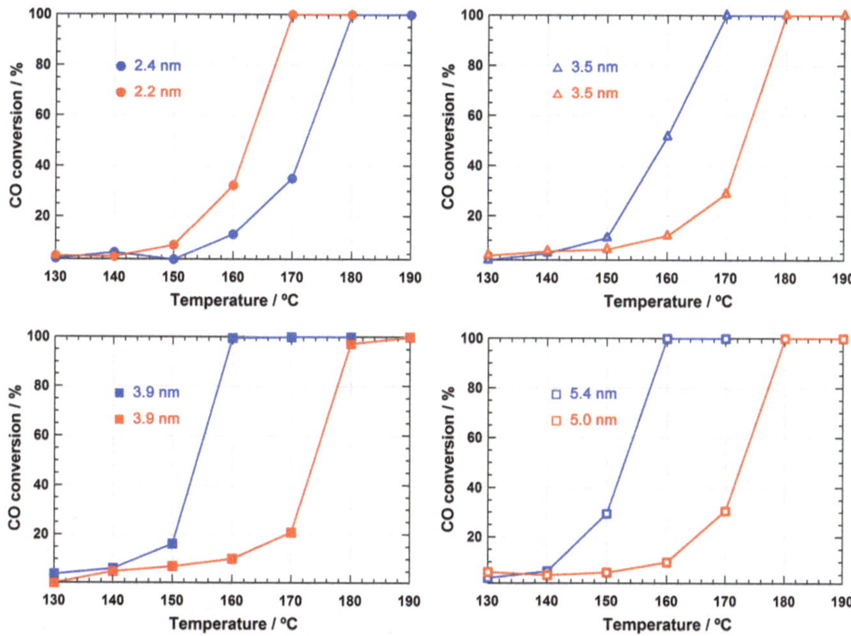

Fig. 4.5 The temperature dependencies of CO conversion in Ru nanoparticles supported on γ-Al_2O_3 (*blue* fcc Ru, *red* hcp Ru)

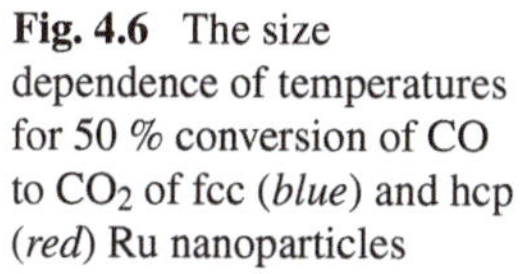

Fig. 4.6 The size dependence of temperatures for 50 % conversion of CO to CO_2 of fcc (*blue*) and hcp (*red*) Ru nanoparticles

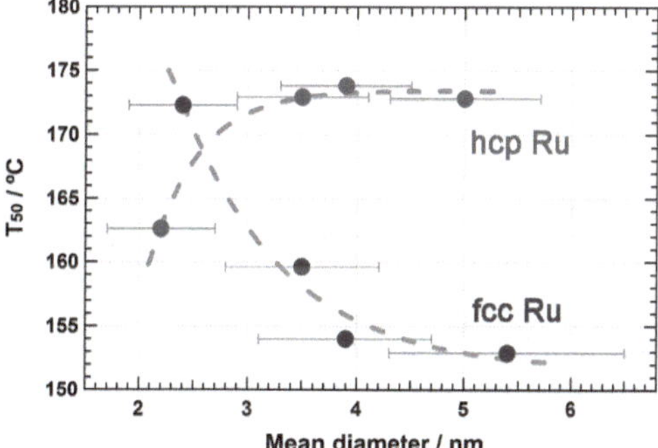

4.4 Conclusion

In summary, the author reports the first synthesis of fcc Ru nanoparticles and their facile synthesis method. While hcp is the only phase in bulk Ru, the fcc phase was obtainable as a nanoparticulate phase. The structure of Ru nanoparticles is controllable by choosing adequate combinations of the Ru precursor and a reducing agent. All of the Ru nanoparticles synthesized with $Ru(acac)_3$ and TEG adopt a fcc structure, while the Ru nanoparticles synthesized with $RuCl_3 \cdot nH_2O$ and EG have a hcp structure. In addition, Ru nanoparticles supported on γ-Al_2O_3 exhibit structure and size-dependent catalytic activity of CO oxidation. From the present results, other metal nanoparticles have a possibility to adopt unknown structures in bulk state, providing unique and valuable properties which are different from the conventional materials.

References

1. Pettifor DG (1977) A physicist's view of the energetics of transition metals. Calphad 1:305–324
2. Saunders N, Miodownik AP, Dinsdale AT (1988) Metastable lattice stabilities for the elements. Calphad 12:351–374
3. Xia H, Parthasarathy G, Luo H, Vohra YK, Ruoff AL (1990) Crystal structures of group IVa metals at ultrahigh pressures. Phys Rev B 42:6736–6738
4. Hanfland M, Syassen K, Christensen NE, Novikov DL (2000) New high-pressure phases of lithium. Nature 408:174–178
5. Shimizu K, Ishikawa H, Takao D, Yagi T, Amaya K (2002) Superconductivity in compressed lithium at 20 K. Nature 419:597–599
6. Errandonea D, Meng Y, Häusermann D, Uchida T (2003) Study of the phase transformations and equation of state of magnesium by synchrotron x-ray diffraction. J Phys Condens Matter 15:1277–1289
7. Ma Y, Eremets M, Oganov AR, Xie Y, Trojan I, Medvedev S, Lyakhov AO, Valle M, Prakapenka V (2009) Transparent dense sodium. Nature 458:182–185
8. Liu Q, Fan C, Zhang R (2009) First-principles study of high-pressure structural phase transitions of magnesium. J Appl Phys 105:123505
9. Tateno S, Hirose K, Ohishi Y, Tatsumi Y (2010) The structure of iron in earth's inner core. Science 330:359–361
10. Stixrude L (2012) Structure of Iron to 1 Gbar and 40,000 K. Phys Rev Lett 108:055505
11. Hrubiak R, Drozd V, Karbasi A, Saxena SK (2012) High P–T phase transitions and P–V–T equation of state of hafnium. J Appl Phys 111:112616
12. Xiong S, Qi W, Huang B, Wang M, Li Z, Liang S (2012) Size-temperature phase diagram of titanium nanosolids. J Phys Chem C 116:237–241
13. Jesser WA, Shneck RZ, Gile WW (2004) Solid-liquid equilibria in nanoparticles of Pb-Bi alloys. Phys Rev B 69:144121
14. Calvo F, Doye JPK (2004) Pressure effects on the structure of nanoclusters. Phys Rev B 69:125414–125416
15. Dong XL, Choi CJ, Kim BK (2002) Chemical synthesis of Co nanoparticles by chemical vapor condensation. Scr Mater 47:857–861
16. Ling T, Xie L, Zhu J, Yu H, Ye H, Yu R, Cheng Z, Liu L, Yang G, Cheng Z, Wang Y, Ma X (2009) Icosahedral face-centered cubic fe nanoparticles: facile synthesis and characterization with aberration-corrected TEM. Nano Lett 9:1572–1576
17. Kim H, Kaufman MJ, Sigmund WM, Jacques D, Andrews R (2003) Observation and formation mechanism of stable face-centered-cubic Fe nanorods in carbon nanotubes. J Mater Res 18:1104–1108
18. Perkas N, Teo J, Shen S, Wang Z, Highfield J, Zhong Z, Gedaken A (2011) Supported Ru catalysts prepared by two sonication-assisted methods for preferential oxidation of CO in H_2. Phys Chem Chem Phys 13:15690–15698
19. Carballo JMG, Yang J, Holmen A, García-Rodríguez S, Rojas S, Ojeda M, Fierro JLG (2011) Catalytic effects of ruthenium particle size on the fischer-tropsch synthesis. J Catal 284:102–108

20. Kim YH, Yim S, Park ED (2012) Selective CO oxidation in a hydrogen-rich stream over Ru/SiO_2 catal. Today 185:143–150
21. Strebel C, Murphy S, Nielsen RM, Nielsen JH, Chorkendorff I (2012) Probing the active sites for CO dissociation on ruthenium nanoparticles. Phys Chem Chem Phys 14:8005–8012
22. Wendt S, Knapp M, Over H (2004) The role of weakly bound on-top oxygen in the catalytic CO oxidation reaction over $RuO_2(110)$. J Am Chem Soc 126:1537–1541
23. Ertl G (2008) Reactions at surfaces: from atoms to complexity (nobel lecture). Angew Chem Int Ed 47:3524–3535
24. Xie X, Li Y, Liu Z, Haruta M, Shen W (2009) Low-temperature oxidation of CO catalysed by Co_3O_4 nanorods. Nature 458:746–749
25. Kaden WE, Wu T, Kunkel WA, Anderson SL (2009) Electronic structure controls reactivity of size-selected Pd clusters adsorbed on TiO_2 surfaces. Science 326:826–829
26. Alayoglu S, Nilekar AU, Mavrikakis M, Eichhorn B (2008) Ru–Pt core–shell nanoparticles for preferential oxidation of carbon monoxide in hydrogen. Nat Mater 7:333–338
27. Qiao B, Wang A, Yang X, Allard LF, Jiang Z, Cui Y, Liu J, Li J, Zhang T (2011) Single-atom catalysis of CO oxidation using Pt_1/FeO_x. Nat Chem 3:634–641
28. Roth C, Benker N, Buhrmester T, Mazurek M, Loster M, Fuess H, Koningsberger DC, Ramaker DE (2005) Determination of O[H] and CO coverage and adsorption sites on PtRu electrodes in an operating PEM fuel cell. J Am Chem Soc 127:14607–14615
29. Jiang H, Liu B, Akita T, Haruta M, Sakurai H, Xu Q (2009) Au@ZIF-8: CO oxidation over gold nanoparticles deposited to metal-organic framework. J Am Chem Soc 131:11302–11303
30. Kim HY, Lee HM, Henkelman G (2012) CO oxidation mechanism on CeO_2-supported Au nanoparticles. J Am Chem Soc 134:1560–1570
31. Kobayashi M, Kai T, Takano N, Shiiki K (1995) The possibility of ferromagnetic BCC ruthenium. J Phys Condens Matter 7:1835–1842
32. Watanabe S, Komine T, Kai T, Shiiki K (2000) First-principle band calculation of ruthenium for various phases. J Magn Magn Mater 220:277–284
33. Lim B, Jiang M, Tao J, Camargo PHC, Zhu Y, Xia Y (2009) Shape-controlled synthesis of Pd nanocrystals in aqueous solutions. Adv Funct Mater 19:189–200
34. González AL, Noguez C, Ortiz GP, Rodríguez-Gattorno G (2005) Optical absorbance of colloidal suspensions of silver polyhedral nanoparticles. J Phys Chem B 109:17512–17517
35. Grass ME, Zhang Y, Butcher DR, Park JY, Li Y, Bluhm H, Bratlie KM, Zhang T, Somorjai GA (2008) A reactive oxide overlayer on rhodium nanoparticles during CO oxidation and its size dependence studied by in situ ambient-pressure x-ray photoelectron spectroscopy. Angew Chem Int Ed 47:8893–8896
36. Haruta M, Tsubota S, Kobayashi T, Kageyama H, Gent MJ, Delmon B (1993) Low-temperature oxidation of CO over gold supported on TiO_2, α-Fe_2O_3, and Co_3O_4. J Catal 144:175–192
37. McCarthy E, Zahradnik J, Kuczynski GC, Carberry JJ (1975) Some unique aspects of CO oxidation on supported Pt. J Catal 39:29–35
38. Over H, Kim YD, Seitsonen AP, Wendt S, Lundgren E, Schmid M, Varga P, Morgante A, Ertl G (2000) Atomic-scale structure and catalytic reactivity of the $RuO_2(110)$ surface. Science 287:1474–1476
39. Gong X, Liu Z, Raval R, Hu P (2004) A systematic study of CO oxidation on metals and metal oxides: density functional theory calculations. J Am Chem Soc 126:8–9
40. Reuter K, Scheffler M (2003) Composition and structure of the $RuO_2(110)$ surface in an O_2 and CO environment: Implications for the catalytic formation of CO_2 Phys. Rev B 68:045407–045411
41. Teranishi T, Kurita R, Miyake M (2000) Shape Control of Pt Nanoparticles. J Inorg Organomet Polym 10:145–156

Chapter 5
Changeover of the Thermodynamic Behavior for Hydrogen Storage in Rh with Increasing Nanoparticle Size

5.1 Introduction

Research into the reaction of hydrogen with metals has attracted much attention because of potential applications as effective hydrogen storage materials, as permeable films, or as catalysts for hydrogenation [1–3]. Nanometer-scale materials as hydrogen storage materials have been investigated recently [4–9] because nanoscale dimensions cause significant changes in material properties from the bulk materials, due to quantum size effects and high surface area to volume ratios [10–12]. For instance, Pd is one of the famous hydrogen storage metals, the total amount of hydrogen absorption of Pd nanoparticles was decrease with decreasing particle size [13]. Although Ag and Rh cannot mix each other at the atomic level in bulk state and both of them do not absorb hydrogen, AgRh solid-solution alloy was obtained as nanoparticles and the alloy exhibited hydrogen storage property [14]. Moreover, bulk Ir and Rh metal do not absorb hydrogen, but nanoscale Ir and Rh do absorb hydrogen with the hydrogen capacity increasing with decreasing particle size [15, 16]. While the hydrogen absorption properties of bulk and sub 10 nm Rh particles have been investigated, there is no report for sizes of ~10 nm. Furthermore, the temperature dependence of hydrogen uptake for Rh nanoparticles has not been reported. Temperature dependence in the reaction of hydrogen with metals gives thermodynamic information which provides a good design guide for hydrogen storage materials. In this chapter, the author presents the hydrogen storage properties of Rh nanoparticles having a diameter of ~10 nm and the temperature dependences of hydrogen absorption by Rh nanoparticles with various particle sizes.

K. Kusada, *Creation of New Metal Nanoparticles and Their Hydrogen-Storage and Catalytic Properties*, Springer Theses, DOI 10.1007/978-4-431-55087-7_5

5.2 Experiment

5.2.1 Synthesis of Rh Nanoparticles Having a Diameter of ~10 nm

Uniformly sized Rh nanoparticles having a diameter of ~10 nm were prepared by a chemical reduction method using $RhCl_3 \cdot 3H_2O$ as the metal precursors and poly(*N*-vinyl-2-pyrrolidone) (PVP) as the stabilizing agent. Ethylene glycol (EG) was used as both the solvent and the reducing agent. $RhCl_3 \cdot 3H_2O$ (5.0 mmol) was dissolved in 20 ml water (solution 1). PVP (5.0 mmol, in terms of monomer unit) was dissolved in 200 ml EG (solution 2). Solution 2 was heated to 196 °C with stirring, and then solution 1 was added into heated solution 2 over 10 min. The mixed solution was maintained at this temperature for 1.5 h. After the reaction was complete, the prepared nanoparticles were separated by centrifugation. The smaller Rh nanoparticles were prepared by following the previous report [16].

5.2.2 Characterizations

Transmission Electron Microscopy (TEM) Analysis

Transmission electron microscopy (TEM) images were recorded on a JEM-200CX or Hitachi HT7700 TEM instruments operated at 200 or 100 kV accelerate voltage, respectively. The samples dispersed with ethanol were drop-cast onto a carbon-coated copper grid and allowed to dry under ambient conditions.

X-Ray Diffraction (XRD) Measurement

X-ray diffraction (XRD) patterns were measured using a Cu K_α radiation source (Bruker D8 Advance diffractometer).

Pressure-Composition (PC) Isotherms Measurement

Pressure-composition (PC) isotherms were measured at hydrogen pressure range between 10^{-5} and ca. 10^{-1} MPa at 303 and 373 K by a volumetric technique using a pressure-composition-temperature apparatus (Suzuki Shokan Co., Ltd).

Solid-State ^{2}H NMR Spectra Measurement

Solid-state ^{2}H NMR spectra were recorded at 303 K in a fixed magnetic field of 94 kOe and a frequency sweep of 60.94–61.94 MHz, using a BRUKER NMR spectrometer. Each sample was evacuated in a glass capillary for several hours at 373 K, and then each sample was sealed into a glass capillary with 86.7 kPa of ^{2}H gas at 303 K.

5.3 Results and Discussion

The TEM images for the prepared particles (Fig. 5.1) showed the formation of uniform nanoparticles with narrow size distribution. The mean diameters of the nanoparticles were determined from the TEM images to be (a) 2.4 ± 0.5, (b) 4.0 ± 0.7, (c) 7.1 ± 1.2, and (d) 10.5 ± 0.8 nm, respectively. Numbers that follow the ± sign represent estimated standard deviations. The mean diameters and distributions were estimated by averaging over 200 particles.

The crystal structure of the prepared particles was investigated by powder XRD (Fig. 5.2). The prepared samples show a diffraction patterns which are consistent with bulk Rh pattern. The broad peaks in the powder XRD supported the formation of a nanomaterial, and the crystal grain size of 10.5 ± 0.8 nm sample was calculated to be 10.5 nm through the Scherrer equation, which corresponds well

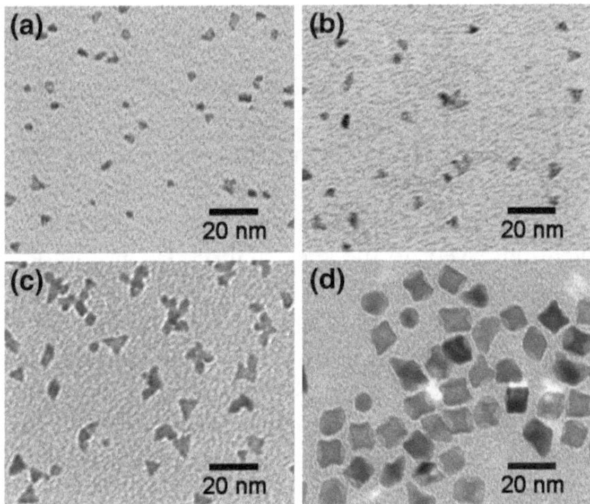

Fig. 5.1 TEM images of **a** 2.4 ± 0.5, **b** 4.0 ± 0.7, **c** 7.1 ± 1.2, and **d** 10.5 ± 0.8 nm

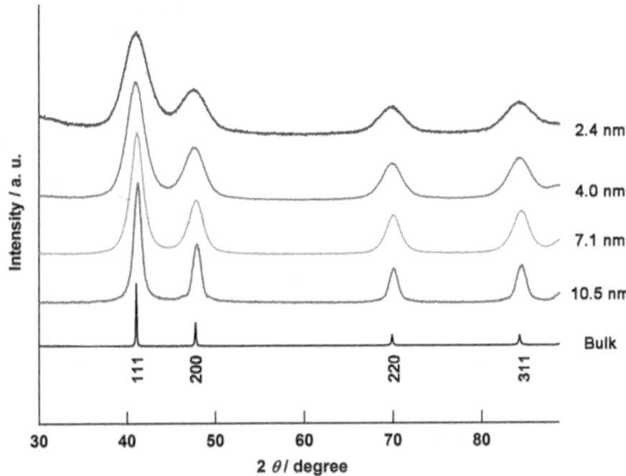

Fig. 5.2 The powder XRD patterns of the prepared nanoparticles with a diameter of 2.4 nm (*blue*), 4.0 nm (*green*), 7.1 nm (*yellow*), 10.5 nm (*red*) and bulk Rh (*black*). The 2θ ranges from 30.0° to 89.0° (color figure online)

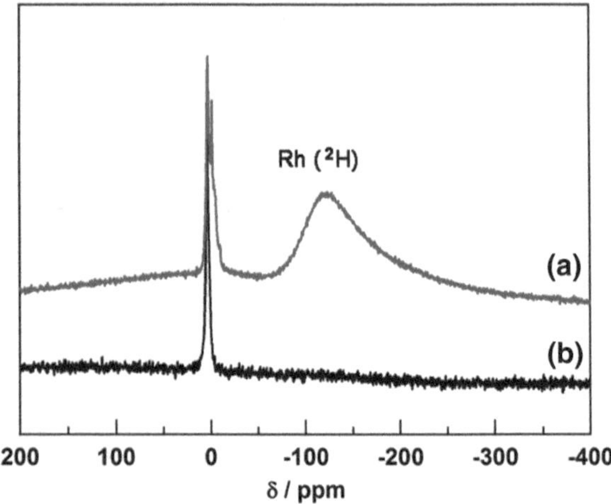

Fig. 5.3 **a** Solid-state ^{2}H NMR spectrum of 10.5 nm Rh nanoparticles. **b** Spectrum for $^2H_2$ gas was also measured for reference. Both spectra were measured at a pressure of 0.087 MPa of $^2H_2$ gas at 303 K

with the TEM image. The XRD data indicates that prepared Rh sample consists of single crystalline nanoparticles.

Solid-state ^{2}H NMR measurement was performed to investigate the state of ^{2}H in the 10.5 nm Rh nanoparticles (Fig. 5.3). In the spectrum of Rh nanoparticles, a broad absorption line at ca. -120 ppm and a sharp line at ca. 0 ppm were observed. In the spectrum of $^2H_2$ gas, only a sharp line at ca. 0 ppm was obtained. On comparison of these spectra, it is reasonable to attribute the sharp line in the spectrum of the Rh nanoparticles to free deuterium gas ($^2H_2$) and the broad component to absorbed deuterium atoms (^{2}H) whose motions are restricted to within the Rh lattice. This result is the same as solid state ^{2}H-NMR previously reported for smaller Rh nanoparticles, so therefore it was found that the larger 10.5 nm Rh nanoparticles also have hydrogen storage capabilities.

To investigate the hydrogen storage properties of the Rh nanoparticles based on particle size, the hydrogen PC isotherms were measured on four samples with particle sizes from 2.4 to 10.5 nm. Hydrogen absorption properties of metals give important information related to the electronic state of the metals [17–22]. As shown in Fig. 5.4, the total amount of hydrogen absorption of prepared 10.5 nm Rh nanoparticles at 0.1 MPa hydrogen pressure at 303 K was 0.067 H/Rh, and this value is less than that of smaller Rh nanoparticles. This result follows previous report that the total amount of hydrogen absorption of Rh nanoparticles decreased with increasing particle size [16]. The most noteworthy outcome is the change in temperature dependence of the hydrogen storage properties. Rh nanoparticles having a diameter 7.1 nm or less can absorb more hydrogen at 303 K than at 373 K. Conversely the Rh nanoparticles having a diameter 10.5 nm can absorb more hydrogen at 373 K than at 303 K. This result shows that the enthalpy of hydrogen absorption for the Rh nanoparticles was changed from exothermic to endothermic as particle size increased.

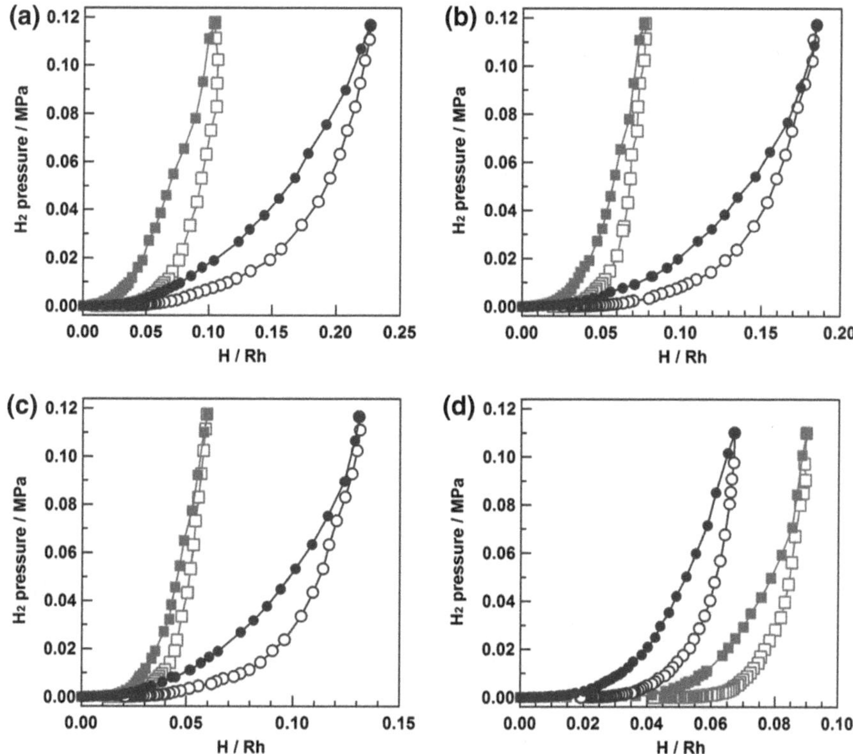

Fig. 5.4 Temperature dependences in PC isotherms of Rh nanoparticles having a diameter of **a** 2.4, **b** 4.0, **c** 7.1, and **d** 10.5 nm, respectively (*blue filled circle* absorption at 303 K, *blue circle* desorption at 303 K, *red filled square* absorption at 373 K, *red square* desorption at 373 K) (color figure online)

To further investigate, the numerical values of thermodynamics parameters of hydrogen absorption in Rh nanoparticles were estimated. At low hydrogen pressures, the hydrogen absorption behaviors of Rh nanoparticles were obeyed Sieverts law,

$$\sqrt{p/p^0} = K_s x,$$

where p is the hydrogen equilibrium pressure, $p^0 = 0.1$ MPa, K_S is Sieverts constant and $x =$ H/M. The temperature dependence of K_S is approximately given by [23]

$$\ln K_S = -\Delta S_S/R + \Delta H_S/RT.$$

The heat of solution (ΔH_S) of hydrogen in Rh nanoparticles which was estimated through above formulas was summarized in Table 5.1. The ΔH_S were increased with increasing the particle size, and the value of ΔH_S changed from minus to plus between the particle size 7.1–10.5 nm.

Table 5.1 The thermodynamic parameters of hydrogen absorption in Rh nanoparticles

Size (nm)	2.4	4.0	7.1	10.5
ΔH_S (kJ/mol H)	−12.3	−4.7	−1.9	6.9

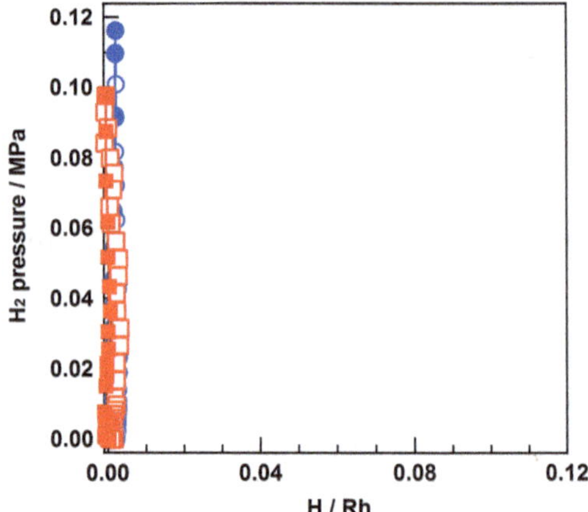

Fig. 5.5 Temperature dependences in PC isotherms of bulk Rh (*blue filled circle* absorption at 303 K, *blue circle* desorption at 303 K, *red filled square* absorption at 373 K, *red square* desorption at 373 K)

This phenomenon might be caused by the change in electronic state of Rh with increasing particle size. The electronic state of 10.5 nm Rh nanoparticles appears to be in an intermediate state between that of bulk and nanoscale particles, which is greatly affected by nanosize effect. From the point of view of electronic state, the 10.5 nm particles have unique hydrogen storage property which is considered as a property in boundary region between bulk and nanosize region. As a reference, the hydrogen PC isotherms of bulk Rh are shown in Fig. 5.5.

5.4 Conclusion

In summary, the author investigated the hydrogen storage properties of Rh in the boundary region between bulk and nanoscale. The enthalpy of hydrogen absorption in Rh nanoparticles was changed from exothermic to endothermic with increasing particle size, and the critical size is between 7 and 10 nm. The thermodynamics of hydrogen storage in metal nanoparticles could be tuned by controlling the particle size. In addition, from the point of view of materials science, materials in this boundary region should be expected to have unique properties (Fig. 5.6).

Fig. 5.6 Graphical abstraction of changeover of the thermodynamic behavior for hydrogen storage in Rh with increasing nanoparticle size

References

1. Holleck GL (1970) Diffusion and solubility of hydrogen in palladium and palladium-silver alloys. J Phys Chem 74:503–511
2. Pundt A, Kirchheim R (2006) Hydrogen in metals: microstructural aspects. Ann Rev Mater Res 36:555–608
3. Boddien A, Mellmann D, Gärtner F, Jackstell R, Junge H, Dyson PJ, Laurenczy G, Ludwig R, Beller M (2011) Efficient dehydrogenation of formic acid using an iron catalyst. Science 333:1733–1736
4. Zaluska A, Zaluski L, Stöm-Olsen JO (1999) Nanocrystalline magnesium for hydrogen storage. J Alloys Compd 288:217–225
5. Kobayashi H, Yamauchi M, Kitagawa H, Kubota Y, Kato K, Takata M (2008) Atomic-level Pd-Pt alloying and largely enhanced hydrogen-storage capacity in bimetallic nanoparticles reconstructed from core/shell structure by a process of hydrogen absorption/desorption. J Am Chem Soc 130:5576–5577
6. Kobayashi H, Yamauchi M, Kitagawa H, Kubota Y, Kato K, Takata M (2008) On the nature of strong hydrogen atom trapping inside Pd nanoparticles. J Am Chem Soc 130:1828–1829
7. Kobayashi H, Yamauchi M, Kitagawa H, Kubota Y, Kato K, Takata M (2008) Hydrogen absorption in the core/shell interface of Pd/Pt nanoparticles. J Am Chem Soc 130:1818–1819
8. Zlotea C, Campesi R, Cuevas F, Leroy E, Dibandjo P, Volkringer C, Loiseau T, Férey G, Latroche M (2010) Pd nanoparticles embedded into a metal-organic framework: synthesis, structural characteristics, and hydrogen sorption properties. J Am Chem Soc 132:2991–2997
9. Zlotea C, Cuevas F, Paul-Boncour V, Leroy E, Dibandjo P, Gadiou R, Vix-Guterl C, Latroche M (2010) Size-dependent hydrogen sorption in ultrasmall Pd clusters embedded in a mesoporous carbon template. J Am Chem Soc 132:7720–7729
10. Kubo R (1962) Electronic properties of metallic fine particles. I. J Phys Soc Japan 17:975–986
11. Henglein A (1989) Small-particle research: physicochemical properties of extremely small colloidal metal and semiconductor particles. Chem Rev 89:1861–1873
12. Roduner E (2006) Size matters: why nanomaterials are different. Chem Soc Rev 35:583–592
13. Yamauchi M, Ikeda R, Kitagawa H, Takata M (2008) Nanosize effects on hydrogen storage in palladium. J Phys Chem C 112:3294–3299
14. Kusada K, Yamauchi M, Kobayashi H, Kitagawa H, Kubota Y (2010) Hydrogen-storage properties of solid-solution alloys of immiscible neighboring elements with Pd. J Am Chem Soc 132:15896–15898
15. Kobayashi H, Yamauchi M, Kitagawa H (2012) Finding hydrogen-storage capability in iridium induced by the nanosize effect. J Am Chem Soc 134:6893–6895
16. Kobayashi H, Morita H, Yamauchi M, Ikeda R, Kitagawa H, Kubota Y, Kato K, Takata M (2011) Nanosize-induced hydrogen storage and capacity control in a non-hydride-forming element: rhodium. J Am Chem Soc 133:11034–11037

17. Wicke E (1984) Electronic structure and properties of hydrides of 3d and 4d metals and intermetallics. J Less Common Met 101:17–33
18. Fazle Kibria AKM, Sakamoto Y (2000) The effect of alloying of palladium with silver and rhodium on the hydrogen solubility, miscibility gap and hysteresis. Int J Hydrogen Energy 25:853–859
19. Papaconstantopoulos DA, Klein BM, Economou EN, Boyer LL (1978) Band structure and superconductivity of PdD_x and PdH_x. Phys Rev B 17:141–150
20. Yamauchi M, Kobayashi H, Kitagawa H (2009) Hydrogen storage mediated by Pd and Pt nanoparticles. ChemPhysChem 10:2566–2576
21. Kobayashi H, Yamauchi M, Ikeda R, Kitagawa H (2009) Atomic-level Pd–Au alloying and controllable hydrogen-absorption properties in size-controlled nanoparticles synthesized by hydrogen reduction. Chem Commun 45:4806–4808
22. Kobayashi H, Morita H, Yamauchi M, Ikeda R, Kitagawa H, Kubota Y, Kato K, Takata M, Toh S, Matsumura S (2012) Nanosize-induced drastic drop in equilibrium hydrogen pressure for hydride formation and structural stabilization in Pd–Rh solid-solution alloys. J Am Chem Soc 134:12390–12393
23. El-Sanabary F, Ramaprabhu S, Weiss A (1993) Thermodynamics of hydrogen dissolved in palladium-rich Pd-Er-Ag(Au, Cu) ternary solid solution alloys. Ber Bunsenges Phys Chem 97:607–617

Curriculum Vitae

Personal Information

Name: Kohei Kusada (Ph.D.)
Sex: Male
Date of Birth: 10/08/1985
Nationality: Japan

Work Experience

2014.04: Assistant Professor at Graduate School of Science, Kyoto University

2013.04–2014.03: Researcher at Catalyst Laboratory, AsahiKASEI Chemicals Corporation

2010.04–2013.03: JSPS Research Fellow (DC1) at Graduate School of Science, Kyoto University

Education

2010.04–2013.03:

Ph.D. at Graduate School of Science, Kyoto University.
Supervisor: Professor Hiroshi Kitagawa

K. Kusada, *Creation of New Metal Nanoparticles and Their Hydrogen-Storage and Catalytic Properties*, Springer Theses, DOI 10.1007/978-4-431-55087-7

2008.04–2010.03:

M.D. at Graduate School of Science, Kyusyu University.
Supervisor: Professor Hiroshi Kitagawa

2004.04–2008.03:

B.D. at Department of chemistry, Faculty of Science, Kyusyu University.
Supervisor: Professor Hiroshi Kitagawa

Zeitfracht Medien GmbH
Ferdinand-Jühlke-Straße 7
99095 Erfurt, Deutschland
produktsicherheit@kolibri360.de